W Pryor

A Treatise on Rupture

Its Causes, Progress and Danger, with and Examination of the....

W Pryor

A Treatise on Rupture
Its Causes, Progress and Danger, with and Examination of the....

ISBN/EAN: 9783337170820

Printed in Europe, USA, Canada, Australia, Japan

Cover: Foto ©berggeist007 / pixelio.de

More available books at **www.hansebooks.com**

A TREATISE

ON

RUPTURE.

ITS CAUSES, PROGRESS AND DANGER, WITH AN EXAMINATION OF
THE CLAIMS OF THE DIFFERENT METHODS BEFORE
THE PUBLIC FOR ITS TREATMENT.

TRUSSES,

THEIR INEFFICIENCY AND THE DANGER RESULTING FROM
THEIR PERSISTENT USE, TOGETHER WITH THE
BEST MEANS OF RELIEF AND CURE.

Also a consideration of that most frequent and distressing form of

DISPLACEMENT,

KNOWN AS "PROLAPSUS UTERI,"

WITH AN EXAMINATION OF THE METHODS OF SUPPORT
FOR ITS RELIEF AND CURE.

*Accompanied by a philosophic exposition of the means of preserving
Health and prolonging Life, in Men, Women and
Children, known as the*

LAWS OF HYGIENE.

By W. PRYOR, M. D.

ST. LOUIS:
CHANCY R. BARNS, PRINTER.
1876.

PREFACE.

The most eminent treatises on the subject of Hernia (misnamed Rupture) are those of Sir Astley Cooper and D. W. Lawrence, F.R.S., both English surgeons of distinguished reputation. These works are voluminous and but illy suited to the popular wants.

As the subject is deserving of the earnest consideration of the profession, and the great body of the laboring population, who are its subjects most generally, I propose to offer, for the consideration of the American public, a monograph, divested of the technical and learned character of those before the public, written in such a simple and clear manner, accompanied by plates illustrative of the subject in all its various forms, so that an interest may be created in this domain of surgery, and the popular mind awakened to the importance of more special attention to this dangerous deformity. If I shall succeed in drawing attention to what seems to me a most interesting but neglected field of labor, I shall be amply repaid for the time spent in collecting materials for this paper.

When I say neglected field of labor, I allude specially to the mechanical devices for the relief and cure of Hernia, now before the public, styled trusses. While their name is legion none have stood the test of time in either relieving or curing —except Dr. J. A. Sherman's Hernial Appliance and Curative Compound

HERNIA.

The habits of excessive toil, to which the human family are doomed, in consequence of their relation to the material universe, which demands employment of violent muscular efforts to procure the necessaries, comforts and conveniences of life, together with the hereditary weakness under which our race labors, through a long line of ancestral transgressions, have developed a physical imperfection, both distressing and dangerous, styled Hernia or Rupture. In a race physically strong we do not believe it could exist. Nor do we believe, even in the present type of feeble and depleted manhood and womanhood, it could exist, if a proper system of physical training were begun by the mother, from the conception of the embryo child to the period of delivery, and thence onward through childhood, till the body was well consolidated or knit, so as to exclude all idea of weakness or want of strength. Such habits, which sound physiological instruction to our sons and daughters would produce, *can alone* banish this physical curse from our race.

We have accordingly devoted much care in the preparation of the most important principles of physiological science, and given them to the public with the hope of calling their most urgent attention to the rules or laws of health applicable to manhood, womanhood and childhood. It is from the *obedience* which follows such knowledge that we can alone hope to banish from our feeble race their hereditary infirmities, and substitute a better, purer and higher standard of physical perfection.

HERNIA AND ITS VARIETIES.

(Epvos, a branch, from its protruding forward.)

A protrusion of any viscus from its proper cavity is called a Hernia. The viscus passes into a cavity called a sac, composed of a portion of the peritoneum, which is pushed before it. In case of Hernia caused by wounds or great violence there is no sac. The parts of the body where the Hernia most frequently appears are the groin, the scrotum, the labia puderdi, the navel, and the upper and fore part of the thigh. If the intestine protrudes, the Hernia is called *enterocele*. If omentum, *epiplocele*. If both, *entero-epiplocele*. When the contents of the Hernia protrude at the external abdominal ring the case is called *Inguinal Hernia* — of which there are two varieties, *direct and oblique*. If the protrusion pass only as low as the *groin* it is called a *Bubono-cele*. But if the protrusion descend into the scrotum it is called *oscheo-cele* or *Scrotal Hernia*; if into the labium, a *Labial Hernia*. The *Crural or Femoral Hernia* is the name given to that which takes place below *Poupart's Ligament*. When the bowels protrude at the navel the case is called *Exomphalos* or Umbilical Hernia. *Ventral* is the epithet given to the swelling when it occurs at any other promiscuous part of the wall of the abdomen. When the protruded bowels lie quietly in the sac, and admit of being readily put back into the abdomen, the case is termed a *Reducible Hernia*. Yet when they suffer no constriction, and cannot be put back, owing to adhesions or their large size, in relation to the aperture through which they have to pass, the Hernia is termed *irreducible*. An *Incarcerated* or *strangulated* Hernia is one which not on only cannot be easily reduced, or cannot be reduced without an operation, but suffers constriction, so that if a piece of intestine be protruded the pressure to which it is subjected stops the passage of its contents toward the anus, or even stops the circulation of the blood in the protruded part, excites inflamation of the bowel, or causes its death, and brings on a train of alarming and often fatal consequences.

CAUSES—*First*, PREDISPOSING ; *Second*, EXCITING.

1st. A large size of the opening at which the bowels are liable to protrude ; a weakness and relaxation of the margins of these apertures ; extraordinary laxity of the peritoneum ; unusually long mesentery or omentum ; and in regard to the abdominal ring, the transverse tendinous fibres *(inter-columnae fi)*, which naturally cross its upper and outer part, are much weaker in some subjects than in others.

Men are much more liable than *women* to Hernia Inguinalis, from the much larger size of the inguinal canal. While in women as there is a larger space for the protrusion of the viscera below Poupart's Ligament, they are more exposed to Femoral Hernia. The distension of the abdomen in pregnancy also gives a disposition to Crural or Umbilical Hernia.

2d. *The exciting causes.*—The grand cause here is the powerful action of the abdominal muscles and diaphragm on the viscera, as jumping, lifting, running, vomiting, straining at stool, parturition, coughing, sneezing, &c., and in people who inhabit mountainous countries. In people having a continued stricture, the effort to cause the urine to pass often brings on Hernia. Hernias from the first cause come on *gradually* and almost imperceptibly. Those from the second are formed *suddenly*, by the immediate action of the exciting cause. In Inguinal Hernia the complaint is often indicated, in the first instance, by a fulness, combined with a sense of weakness about the abdominal ring. The swelling is increased by any action of the respiratory muscles, and disappears on pressure, and in the recumbent position of the body. It gradually finds its way through the tendon of the ex. abdominal muscle into the groin, and afterward into the scrotum. When a Hernia occurs suddenly there is a sensation of something giving way at the part, and with pain.   (Lawrence.)

Scarpa says that Martin, Benevoli, Bendal and Morgagin consider a relaxation and elongation of the mesentery as the principal cause of Hernia in general, and of Inguinal Hernia in particular.

In a *healthy abdomen* there are two opposite forces which reciprocally balance each other. *One* the pressure of the viscera against the abdominal parieties. The *other* the reaction of these same parieties upon the viscera which they contain. If these two forces were in perfect equilibrium in all individuals, and under all circumstances in life, we could not be in the least subject to Hernia. But there are certain points of the abdominal parieties which naturally present much less resistance than others, and which react with much less power against the pressure made from within outwards by the abdominal viscera. Such is particularly the part which extends from the pubis to the anterior superior spinous process of the Ilium.

Hernia is more frequent on the right than the left side of the body.

SYMPTOMS OF REDUCIBLE HERNIA.

An indolent tumor at some point of the abdomen, most frequently descending out of the abdominal ring, or from just below Poupart's Ligament, or else out of the navel. Swelling often sudden — changes its size — pressure makes it smaller — increases after a meal ; also in standing up.

PROGNOSIS

Depends on age, constitution, date of disease, strictured or not, inflammation present, symptoms returnable or not. In the child more favorable than in the adult. In very old people symptoms not so rapid, owing to the laxity of their frames and more languid circulation, and the age of their ruptures, causing a dilated passage. If the tumor be small, the prognosis more serious. Hernias which arise spontaneously, and merely from predisposing causes, seldom become strangulated, because the parts concerned are weak and relaxed. The aperture is generally very small in Femoral Hernia. This rupture in men, and the Bubono-cele in women have a particularly narrow entrance. On the same grounds, Femoral, Inguinal and Umbilical ruptures are more dangerous than the Ventral, Peritoneal or Vaginal kind.

TREATMENT OF REDUCIBLE HERNIA

Perfect reduction at once, and the application of a suitable support, that keeps the gut within its natural channel. Else, *time is lost and the patient deceived to his ruin.* Must not be careless. As a precautionary measure the support should be continued some time after a cure is effected. Hernias become incurable when neglected, and remain down. They prohibit all active exertion, and prevent, in the male, the act of copulation.

ANATOMY OF PARTS CONCERNED IN FEMORAL AND INGUINAL HERNIA.

The abdomen is covered principally by five pairs of muscles with their tendons. On each side we have the External Oblique, the Internal Oblique, the Transversalis, the Rectus and the Pyramidalis. The first three only are concerned in the Hernias under consideration. The external abdominal muscles arise from the eight inferior ribs on each side, descend towards the lower part of the abdomen, and end in an expanded tendon, which covers the whole of the hypogastric and part of the umbilical regions. Man being destined to walk erect, a firm, inelastic, strong membrane is necessary to protect his abdominal viscera from accidents to which his erect attitude would subject him. In quadrupeds, destined to the horizontal position, the weight and pressure of the viscera are diffused over the whole of the abdominal walls ; *here* no such tendinious expansion is needed, and they have no Hernias. But in man, by the constant action of his diaphragm and abdominal muscles, his abdominal viscera are thrust towards the lower part of the belly, and muscular fibre would prove a feeble barrier, and Hernia would be the invariable result of muscular exertion. And had there not been *two* openings in this aponeurotic shield, man would have been ever exempt from Hernia. Now, on the lower part of the tendon of each muscle, a little above and to the outer side of the symphysis pubis, an opening exists called the abdominal ring, formed for the passage of the spermatic cord to the testicle in the male, and the round ligament of the uterus in the female. The abdominal rings pass obliquely upward and outward, and are rather of a triangular form. Covering the external oblique muscle a dense *fascia* is found under the integuments. This

accompanies the spermatic cord in its descent to the scrotum.
It is known as Superficial Fascia. It acts as a covering to
both Inguinal and Femoral Hernia.

OF THE ORDER IN WHICH THE VARIOUS METHODS OF REDUCTION HERETOFORE AND AT PRESENT USED SHOULD BE RESORTED TO, AND OF THE TIME WHEN THE VARIOUS OPERATIONS SHOULD NO LONGER BE DELAYED.

It is a matter of the greatest importance that *time should
not be wasted* in the trial of methods of reduction. The danger is imminent, for gangrene sometimes speedily comes on
and the patient loses his life.

Baron Larcey records a case, which took place in the French
army in Egypt, when from the period of inception to the
death of the soldier only two hours elapsed.

The Taxis is among the first methods to be tried, and Sir
Astley Cooper recommends the duration to be from a quarter
to half an hour. If after this period no relief has taken
place, then some of the following methods of reduction may
be resorted to by the attending surgeon, and after these the
Taxis again may be tried. In very acute cases when the
patient is strong and young, bleed first, Desault and Scarpa
say, and then put the patient in a warm bath *before* the Taxis
is tried. *Time*, however, must not be wasted in getting the
bath. If at hand, use it, and use the Taxis while the patient
is lying in the bath. Next in order comes the bag of ice or a
solution of nitre and muriate of ammonia applied to the
ruptured part — and then if the patient is not too aged nor
too debilitated, the injection of the infusion of tobacco as

recommended under this head. Four hours, Sir Astley
Cooper says, should be consumed in the use of the clyster
and cold.

As to the time when the operation should be performed, the
general sentiment of the profession seems to be, within the
period of ten hours after the occurrence of the strangulation.
A longer time places the life of the patient in jeopardy.
Better too early than too late. The urgency of the symptoms
must determine the surgeon. (After the Taxis has failed in
the warm bath, the course *now* is to place the patient immediately under the influence of *chloroform* ; then try the Taxis
— and if a failure, to operate without delay.)

DISEASES WITH WHICH HERNIA MAY BE CON-
FOUNDED, AND HOW TO DISCRIMINATE
BETWEEN THEM.

These are Varicocele, Bubo, Hydrocele and Hernia Humoralis, or inflamed testicle. Variocele is an enlargement of the
spermatic veins. From Hernia we may know it thus : Put
the patient in a horizontal position ; press upon the scrotum
and empty the swelling; then put the fingers firmly on the
abdominal ring ; a Hernia will not return. A Bubo declares
itself by being circumscribed, incompressible and hard, and
by its situation, and its freedom from all connection with the
spermatic process. These indicia will point it out, if recent.
When of long standing, and suppuration has taken place, the
fluctuation of the parts will readily point it out from a piece
of intestine or omentum.

From Hydrocele we may readily distinguish it by the following signs : Hydrocele is void of pain on handling—fluctuation
of its contents obvious — formation gradual — begins below

and proceeds upwards, and is not increased by coughing or
sneezing. It is also transparent. From Hernia Humoralis it
is known, by the pain in the testicle.

TREATMENT OF IRREDUCIBLE HERNIA, FREE FROM STRANGULATION AND UNATTENDED WITH TROUBLESOME AND DANGEROUS SYMPTOMS.

The best writers on Ruptures ascribe the inability to reduce
to three causes: 1st. Great size in the contents. 2d. Al-
teration of form or texture. 3d. To adhesions with each
other or with their containing bag, and perhaps sometimes by
transverse, membranous bands within the sac.

Irreducible Herniæ, from either of the above causes, are
not capable of being relieved by surgery. The only appliance
holding forth comfort to the patient is the suspensory bandage, which lessens the inconvenience of the weight of the
tumor and checks the increase of the swelling by means of
pressure.

If the Hernia is omental only, Sir Astley Cooper recommends a spring truss, for the purpose of checking its increase
and subsequent descent, and should pain result, the truss
must be instantly abandoned. Persons afflicted with Irreducible Hernia should be particularly careful not to attempt
feats of strength beyond their capacity, nor aim at feats of
agility, and keep the loaded scrotum out of the way of all
harm. The bowels must be carefully looked too, and cos-
tiveness avoided. The danger here is inflammation, (which
may prove fatal), or a stricture, which is equally hazardous.

SYMPTOMS AND TREATMENT OF A STRANGULATED OR INCARCERATED HERNIA.— Means to be Tried Before an Operation.

When the protruded parts cannot be immediately replaced, and there is pain, and the patient is prevented from going to stool, this is termed "An Incarcerated Hernia, a Strangulated Hernia, or a Hernia with Stricture. This situation is very dangerous, demanding immediate assistance." A stricture on the prolapsed part of the gut, by the aperture or canal through which it passes, is the immediate cause of all the dangerous sypmtoms, and, of course, the release of the intestine from this stricture is the only thing that can give relief. The bowel must be returned back into the abdomen, or the parts divided, which produce the stricture, and it is needless to add, the first is the most desirable when practicable. The various plans to accomplish this are as follows :

1. THE TAXIS.—By which we mean the operation of reducing the Hernia with the hand. The body should be placed on an inclined plane, with the thighs bent towards the trunk. This relaxes the fascia of the thigh, and the aperture through which the Hernia passes. If the Hernia be an Oblique Inguinal one the pressure made on the tumor must be directed upwards and outwards, along the course of the spermatic cord, and be continued from a quarter to half an hour. If Femoral, as it passes downwards and forwards, the pressure must be made, first backwards and then upwards. If Umbilical or Ventral it is to be made straight backward It must ever be remembered that no violence is to be used. This, besides being unavailing, aggravates the inflamed parts, and has sometimes burst the intestine.

While bending the thigh care must be taken to rotate it inwards, as this greatly relaxes the femoral fascia and tendon of the external obliqe muscle.

Mr. Balfour suggests the plan of gently pulling the intestine downwards, or a little further out of the ring, previous to the attempt at reduction. Should the Taxis not succeed at first, it may do so after the warm bath, bleeding, or cold applications.

Small Herniæ, having the closest stricture, are the most difficult to reduce, and for a like reason, crural ruptures yield not so often to the Taxis as Inguinal Herniæ, in the male. The longer the viscera have been down, the less likely the Taxis to succeed, owing to the formation of adhesions.

Mr. W. Lawrence observes, When the rupture becomes painful we are not justified in continuing our efforts by Taxis. This only tends to increase the inflammation and accelerate the approach of gangrene.

At this period the operation is required, and should be performed without delay.

In acute strangulations both Scarpa and Guthrie recommend bleeding first in connection with the warm bath, warm fomentations and emollient clysters, before the Taxis.

In chronic cases, on the contrary, of old, large Herniæ, in feeble or aged persons, the case is different. Here we must support the strength. All weakening agents must be avoided. They produce a general atony and might bring on gangrene of the intestine. Here we must use cold in connection with the Taxis.

2. BLEEDING.—The most eminent surgeons of the past century favor this method of aiding reduction by Taxis. Its control over inflammation — the weakness it temporarily produces, and often the faintness it induces, are the reasons

assigned for its advocacy. From fourteen to twenty ounces of blood are recommended to be drawn. (These remarks were considered true before the introduction of chloroform. Now "bleeding is seldom resorted to," says Sir Astley Cooper, "and never to produce merely faintness and loss of muscular power.")

3. PURGATIVE MEDICINES.—These have been laid aside as not only inefficient, but dangerous.

4. WARM BATH.—This, taken in the horizontal position, is beneficial.

5. COLD BATH AND COLD APPLICATIONS.—The best form in which cold can be used is the pounded ice tied up in a rubber sack and applied to the part. "If, after this trial," says Sir Astley Cooper, "the symptoms are mitigated, the remedy may be continued, but should the tumor continue to resist our efforts at reduction, it may be abandoned."

6. OPIATES.—Not generally successful. When given they should be used in large doses.

7. TOBACCO CLYSTERS.—Next to the operation, these are considered the most efficient in bringing about a reduction of the Hernia. For this purpose about one ounce of the plant is placed, for ten minutes, in a pint of boiling water. Inject one-half of this at first. The rest may be used when it appears the tobacco does not operate with the extraordinary quickness and force which it does on a few constitutions. (Now chloroform takes the place of tobacco.)

THE OPERATION FOR INGUINAL HERNIA.

The aim of this is to free the strangulated intestine from its stricture and replace it in the cavity of the abdomen. A table is selected three feet six inches in height, the body of the patient placed on it in a horizontal position, the shoulders raised, and his legs, from the knee, hanging down over the edge, with the thighs bent in order to relax the muscles of the abdomen. Previous to this the bladder must be emptied and the hair shaved from the hernial side. "The surgeon stands before the patient, between his thighs; then grasping the tumor in his left hand, and with a common scalpel in the right, begins his incision opposite the upper part of the abdominal ring, about the middle of the sac, two-thirds down the tumor." The skin and cellular membrane are thus divided over the sac, also the pudendal artery, which crosses the sac near the ring, and causes some hemmorrhage, which is easily relieved by compression, or the ligature, if the amount of bleeding demands it. The Fascia, which passes off from the external oblique muscle, and which lies beneath the skin and is exposed by this first incision, forms the first and thickest covering of the sac. The next step is to cut, through the middle of this fascia, and then introduce a director and carry it upward to within an inch of the abdominal ring, then divide the fascia upon it, then turn the director DOWNWARD and make a like division of the fascia to the bottom of the tumor. We are now at the second covering of the Hernial sac, the cremaster muscle, which must be divided, aided by the director, just like the fascia, and we have now a perfect exposure of the sac. Should the hernia prove Intestinal, without any adhesions to the sac, a fluctuation may be felt at its anterior inferior part, and then the tumor is grasped and the fluid pressed forward. The surgeon now raises the sac by pinching the cellular membrane which adheres to the anterior inferior part of it, and with the sac thus raised and separated from the intestine, he places the edge of his knife horizontally and cuts a small hole, just large enough to admit the blunt end of a probe or director on which the sac is further divided upwards to the abdominal ring, and downwards far enough to view its contents. Fluid now escapes from the sac, and the contents of the Hernia appear.

The next step is to divide the stricture, which the surgeon accurately explores with his finger, carried into the sac. The stricture may be formed, 1st, at the external abdominal ring; 2d, an inch and a half to two inches above the ring, lying outwards towards the spinous process of the Ilium; 3d, in the mouth of the sac.

Should the stricture be owing to the pressure of the columns which form the ring, the surgeon introduces his finger into the sac as far as the stricture, then conveys the probe pointed bistoury on the fore part of the sac and carries it within the ring; cuts through it on a director upwards, opposite the middle of the sac, enlarging the aperture just so much as may be neccessary to allow the tumor to pass into the abdomen.

DEFINITION OF TERMS USED IN THIS TREATISE.

Scrotum—The integuments that cover the testicles.

Peritoneum—A serous membrame which lines the abdominal cavity.

Spermatic Cord—The cord by which the testicle is suspended.

Round Ligaments — Folds of the peritoneum passing horizontally from the uterus to the sides of the pubis.

Poupart's Ligament — A broad band of tough, flexible membrane, extending from the anterior superior spine of the Ilium to the spine of the os pubis.

Umbilical Cord — Commonly known as the navel string. The cord that binds the child in the womb to its mother.

Perinaeum — The space between the anus and the genital organs.

DIRECT INGUINAL HERNIA.

This species of Hernia is distinguished from the *Indirect*, or oblique, by the direction it takes in reference to the spermatic cord. The latter follows the course of this cord, and has, therefore, the epigastric artery on its *inside*, at the orifice where it first quits the abdomen. Sometimes, however, the *hernial sac* comes out nearer the *pubis* than usual, passing down immediately behind the abdominal ring, with the *epigastric artery* on the *outside*, and the *Hernia* on the *inside*. It is produced, Sir A. Cooper says, either by weakness, or violence done to the tendon of the external oblique Muscle, at Poupart's ligament, which causes a protrusion to to take place immediately behind the ring. The sac runs as follows : 1st, between the fibres of the tendon of the transversalis, one inch above the ring—carrying the fascia transversalis before it ; then proceeds under the tendon of the lower edge of the internal oblique ; with the *epigastric artery* running on the *outside* of the sac, then escape from the ring, with the spermatic cord on the outside, and covered with the fascia of the external oblique.

It is found in three different conditions : 1st, Reducible ; 2d, Irreducible ; 3d, Strangulated.

The first demands a proper support ; the second, to be treated as the common Inguinal Hernia, before described ; the third : in attempting to reduce it, the tumor must be grasped with one hand, and the fingers of the other placed at the abdominal ring, to *knead* the hernia, at the same time *pressing upwards and inwards instead of upwards and outwards*.

The operation for the removal of stricture of this variety of Hernia, is thus performed: The *middle* of the tumor is selected for the incision, and the integuments are then cut through from its upper to its lower part, always following in your incision the direction of the tumor; the fascia are then divided down to the lower extremity of it. The hernial sac is now exposed, and then opened down to the lower part of it; the finger of the surgeon is now passed into the sac, and feels for the stricture; the blunt-pointed bistoury is then introduced between the ring and the sac, slitting the ring directly upwards, till the opening is large enough to admit of the return of the Hernia.

INGUINAL HERNIA—in the Female.

The fundus of the uterus, in woman, is sustained by a ligament, which is attached to the fat and integuments of the pubis. Its course is similar to that of the spermatic cord in the male. It escapes from the abdomen half way between the spinous process of the ilium and the symphisis pubis, under the edges of the transversalis muscle and internal oblique, above Poupart's ligament, leaving the epigastric artery on the inside.

Before the round ligament escapes through the ring, to be attached to the pubis, its course is oblique, like that of the spermatic cord, making the inner ring two inches to the upper and outer part of the abdominal ring. As the ligament is narrower than the spermatic cord, the opening through which it passes in man is larger than in woman; Inguinal Hernia is therefore more frequent in the male than in the female. The

Hernia appears as a tumor, about the size of a pigeon's egg, in the upper part of the labium pudendi. Causes and symptoms the same in the female as in the male. If the tumor is *above* the edge of Poupart's ligament, the Hernia is inguinal; if *below*, femoral. It is like the other forms of Hernia, either reducible, irreducible or strangulated. To the first form, a proper support is the required remedy; to the second, the application of a common T bandage must be made; to the third, taxis must be first used as before described; if this fails, then, in the order and manner before mentioned, we must use the tobacco clyster, ice, &c. If compelled to operate, it is done thus : Place the female in the same position as the male, then make an incision at the abdominal ring, and continue it down to the lower part of the tumor ; the fascia is thus exposed, cut through this and the sac is exposed ; make a small incision in this and examine the contents. Now introduce a bistoury between the sac and the ring. The stricture is then to be dilated and treated as in the male. If the contents of the sac are not mortified they are to be put back into the abdomen, and parts closed as in any other wound.

CONGENITAL HERNIA,— or Hernia Tunica Vaginalis.

The protruded parts in this case are not, as in Inguinal Hernia, contained in a peritoneal sac, but in the bag of the tunica vaginalis of the testicle, and in immediate contact with the testis ; and is thus liable to be confounded with common hernia. Marks of distinction : 1st, In common hernia the testicle is below the sac ; 2d, Common hernia is gradual in its formation ; congenital, sudden and complete.

To be treated in its three forms of reducible, irreducible and strangulated, as the common hernia.

CRURAL HERNIA.

Symptoms.—Pain, on straightening the thigh, extending up to the stomach, causing nausea; on examining the thigh an absorbent gland is felt in the groin more distinctly than in the other, and causes uneasiness on pressure.

The same general symptoms distinguish it as inguinal hernia. When erect it appears, when recumbent it disappears; when the patient coughs it expands; is elastic, and uniform to the touch when intestine protrudes, and gives a gurgling sound when returned into the abdomen. Sometimes mistaken for Inguinal Hernia. Distinction: The neck of inguinal hernia is above the tuberosity of the pubes; crural, below it and to its outer side.

I.—REDUCIBLE.—Taxis as follows: Patient placed horizontaly, thighs bent, and the knees thrown inwards; the surgeon over the body of the patient, places both thumbs on the tumor and presses gently *directly downwards*, and not towards the abdomen.

This pressure is kept up steadily for some minutes, till the tumor is even with the line of the crural arch, *then press towards* the abdomen, and the tumor will return into it. Dr. Sherman's Hernial Appliance is the best remedy.

II.—IRREDUCIBLE.—*Pressure*, recommended by Sir A. Cooper, as the remedy. He prefers a hollow truss.

III.—STRANGULATED.—The same efforts at reduction as in the other kinds of Hernia, before the operation. Sir A. Cooper has known the operation performed successfully on the eighth day after strangulation.

Operation—Thus: Incision began opposite the middle of the tumor, and continued to its fundus. It is made downwards at right angles with the crural arch ; another is made at right angles with the first, in the direction of the longer axis of the tumor ; thus the two form a letter T reversed. This exposes the tumor. The fore part of the sac is then opened, the finger is then passed into the sac and the fascia divided upwards and below ; this opens the hernial sack below the crural arch. The finger is then passed into the sac and the sheath divided upwards, and the tendon cut obliquely inwards and upwards. If the intestine is sound, it is replaced in the abdomen.

UMBILICAL HERNIA.

This passes through an opening at the navel. A small tumor first appears at the navel the size of the tip of the finger, easily returned. If nothing is done to check its growth, it becomes of enormous size, passing downwards towards the pubis. If intestinal, very dangerous. If no bandage is worn, pain is excessive, bowels irregular, and the tumor inflames and swells to a great size. Very frequent in young subjects. Pregnancy is the most frequent cause in the adult female.

I.—Reducible and Irreducicle Umbilical Hernia.— In attempting to reduce, first elevate the shoulders, and raise the pelvis, then bring the thighs at right angles with the body. The surgeon then grasps the tumor with his hand and presses it upwards, and kneads it with his finger and thumb of the other hand. This will replace it. A support must now be applied.

The irreducible grows to an enormous size. A proper support recommended as a measure of relief; when not admissible, a suitable bandage.

II.— STRANGULATED UMBILICAL HERNIA.— The earliest period at which death results here is seventeen and a half hours after the occurrence of symptoms. Indiscreet eating the most frequent cause. After taxis has failed, Sir A. Cooper has found the tobacco clyster the most successful.

Should an operation be necessary, it is performed thus: Make a crural incision through the integuments, which exposes the hernial sac; dilate the opening and the intestine will be seen by drawing the omentum upwards. The stricture is next divided by passing the finger to the umbilical ring, and dilating it upwards toward the sternum.

VENTRAL HERNIA.

It differs from umbilical only in its seat. Any protrusion through the walls of the abdomen, except at the umbilicus or the abdominal rings, is named ventral hernia.

It is rare, and usually presents itself half way between the umbilicus and the ensiform Cartilage in the linea alba. When reducible, after reduction, apply a suitable support. If irreducible, use a proper support.

If strangulated, first try the reduction by taxis, &c.; when they fail, operate as in umbilical hernia, by beginning the incision with the T shape.

PUDENDAL HERNIA.

This is found in the middle of the external labium pudendi, about the size of a pigeon's egg, and shaped like a pear. Sir A. Cooper recommends a truss, employed for the relief of prolapsus, and to be used when this form of hernia is irreducible

If irreducible, the only relief is from the constant support of a bandage. When strangulated, try reduction first; if it fails, operate thus: Make an incision into the labium to expose the lower part of the tumor, then open the hernial sac, and expose the intestine. Now pass a bistoury up the sac, and guided by the finger, introduced into the vagina, divide the sac inwards toward the vaginia. Be sure to empty the bladder before operating or attempting to reduce.

VAGINAL HERNIA.

The space between the uterus and rectum is the point where this hernia is found. "When reducible, a pessary is the remedy."—[Sir Astley Cooper.

PERINEAL HERNIA.

The protrusion here occurs between the bladder and rectum in men, and between the rectum and vaginia in women.

"When reducible, a pessary is to be worn."—[Sir Astley Cooper.

THYROIDEAL HERNIA; or Hernia Foraminis Ovalis.

The protrusion occurs in this case through the opening, in the anterior and upper part of the thyroid foramen of the pelvis. When reducible, use a proper support. When strangulated, the case, Sir A. Cooper says, is hopeless.

VESICLE HERNIA; or Hernia of the Bladder.

This is very rare. Sir A. Cooper has seen only two cases in the living subject. The remedy is a suitable truss, and constant attention to the complete discharge of the urine.

ISCHIATIC HERNIA.

This is exceedingly rare, and has no external tumor.

PHRENIC HERNIA.

The protrusion here is through the diaphragm. The bowels are interrupted, vomiting takes place, extreme abdominal pain, and great pressure on the lungs, causing cough and difficult breathing. No remedy but quiet.

MESENTERIC AND MESOCOLIC HERNIA.

Here the intestine is forced into the mesentery, which is formed of two layers of the peritoneum. Usually caused by a blow on the abdomen. Not fatal.

PART SECOND.

—

THE TRUSS.

THE TRUSS.

This is an instrument designed for the alleviation or cure of that distressing and dangerous physical infirmity, of which we have endeavored to furnish a brief but accurate account in the preceding dissertation. The extent to which this alarming defect exists is stated to be one in every ten of our population. That it should exist at all, is only one of the many evidences of the physical deterioration of our race—a deterioration which should not exist, and could not, did our people justly appreciate the value of physiological science as an indispensable element in all sound systems of education, for both males and females. We repeat what we have so often said, and cannot say again and again with too much emphasis, that the study of our bodies and brains should constitute the first principal element in all systems of education. It is through these that all our joys and all our woes, while tenants of this earth, do come ; and consequently from an obedience to the laws which govern them that all excellence results. When the Truss was introduced, we have no accurate data to determine. We know it could not have been

> "——when the world was in its prime,
> When the first stars had just begun,
> Their race of glory ; and young Time
> Told his first birth-days by the sun."

for this was in the infancy of creation, when physical infirmities were unknown to the race ; a period, according to the book of Genesis, comprising the first 2369 years of human history. "Over all that expanse of time—for more than one-third part the duration of the human race—not a single instance is recorded of a child born blind, or deaf, or dumb, or idiotic, or malformed in any way. But as early as the third generation from Adam, polygamy began, and intermarriages were all along the order of the day. About this period we may suppose ruptures began, and men then began to devise ways and means for their relief—for the idea of a cure does not seem to have entered, then, into the minds of men. A bandage was perhaps first used, but this proving only temporarily beneficial, a more permanent and safer method must be devised,. An iron spring with pads at each end, was the next device perhaps, but a trial of this soon satisfied them that its pain-causing power was unendurable, and it was soon abandoned for something more appropriate. And here comes in perhaps the elastic spring—a remedy now so universally applied, but which we hope to show has been as universally condemned, because both mischievous and dangerous to all who use it; and we expect to prove to the satisfaction of all intelligent minds, that from the danger and misapplication of this spring, which is the leading feature of nearly all of the trusses before the public, has grown the wholesale condemnation of these instruments of torture, discomfort and danger, and caused the general dissatisfaction prevailing, and a looking after something more rational, and better adapted to relief and ultimate cure. We honestly believe such a device is now offered to the public acceptance in the Hernial Appliance of Dr. J. A. Sherman, of N. Y.; a mechanical contrivance, which for beauty, simplicity, comfort, and adaptation to the end to be accomplished, has never been surpassed, nay, not equalled by anything in Europe or America.

Before entering in detail upon the special excellencies of this wonderful contrivance, we will pass in review the course

usually pursued py persons when they find themselves ruptured; next, the anatomical structure of the parts concerned
in hernia, in order to an intelligent comprehension of the
mode of cure. We will then examine in detail the trusses
before the public, and show by comparison, and a critical
analysis of their defects and merits, why Dr. Sherman's
appliances are more deserving of popular support; and
conclude with' the results of a diligent personal inquiry of
some of the principal sufferers, both here and elsewhere, of
the effects of these appliances, in conjunction with his medicinal remedy, on themselves.

Inguinal Hernia is the 'descent of some portion of the
abdominal viscera along the line of the spermatic cord, to a
greater or less extent, to the scrotum.

When a person finds himself the subject of rupture, being
ignorant of its symptoms and the proper plan of treatment, he
usually applies to a surgeon, an apothecary or druggist, or to
an instrument maker—for a truss. After an examination, the
case is generally treated lightly; he is told, perhaps, 'tis
nothing serious; to get a truss and wear it for a few days, and
all will then be well. The inquirer, full of hope and confidence, does as he is bid, and proceeds to an apothecary, and
has an instrument put on—of course, the one selected is in
harmony with the vendor's *prejudices*, for little is scientifically
known about these mischevious devices, which at best are only
temporary palliatives. What fearful consequences are in store
for the deluded man! In its use his hopes change to fears. The
encouragement of his physiciam dies out midst sadness and
suffering. Truss after truss is put on,—pain and displacement
follow,—until injured, disheartened, despairing, he finds himself ruined, physically, and his path of life before him dark
and disheartening.

The affliction, be it remembered, is deceptive in its nature and prognosis. Slight and harmless at first, it awakens no suspicion of its mischief; consequently many go for years with such little inconvenience that they become negligent, and when they least expect it, are suddenly taken with the symptoms and dangers of strangulated rupture.

Trace the origin of this trouble, and you find its fountain head may be sought in the imperfect education of his physician and adviser, and in himself. *In his physician*, because so much engrossed with his patients, he has bestowed little attention on Rupture as a specialty, and *still less* on the most *available mechanical appliances and remedies for its relief and cure*. *In himself*, that, in common with the members of society generally, his education has never been directed to the physical infirmities to which human beings are subject.

We design this treatise to cover this practical and important field of inquiry, by laying before each one, such a simple and intelligent exposition of the nature of Hernia, its symptoms and treatment, that there can be no misapprehension, after its diligent perusal, of the sufferings and dangers to which every one is liable, (especially the weak,) who is so unfortunate as to be afflicted with Hernia.

"It is known as *Oblique Inguinal Hernia;* when entering the internal abdominal ring it traverses the inguinal canal, emerges at the external abdominal ring, and descends to the scrotum in close association with the spermatic cord; as Direct Inguinal Hernia, when, breaking through the external abdominal ring, it discends to the scrotum, but not in the direct and close association with the spermatic cord, as in the other case."

Oblique Inguinal Hernia is by far the most frequent occurrence; so much so, indeed, as to render the idea of a Direct Inguinal Hernia an exceedingly hypothetical affair in the minds of some very good surgeons; and when we reflect upon the nature of the structures to be broken up to produce a Direct Hernia, and the comparatively easy egress of the viscus by the oblique route, we are almost led to the belief that a Direct Hernia exists rather in the fancy of the surgeon than in the phenomena of surgical accidents.

Observe! A Direct Inguinal Hernia is when a viscus breaks through the walls of the abdomen at the abdominal ring. What is the internal ring? Is it an opening that weakens the walls of the abdomen at that point? No: The internal ring is formed by the testicle which has left its bed beside the kidneys, pushing a fold of peritoneum before it in its descent through the channel that nature had provided for its pathway to the scrotum ; and a viscus seeking escape from the cavity of the abdomen will scarcely form a *new* route when one has already been opened for it. Hence Oblique Inguinal Hernia is by far the most frequent, if not the only form of Inguinal Hernia.

A moment's consideration of the course traveled by an Oblique Inguinal Hernia will serve not only to make this idea apparent, but to elucidate the plan of treatment, having in view its radical cure by the use of the truss.

We have seen the Hernia enter the abdominal ring. We have learned how that ring is formed. Next it traverses the Inguinal canal. How is that canal formed, and what are its boundaries?

The floor and external wall of the Inguinal canal are formed by the tendon of the external oblique muscle, that tendon being reflected on itself so as to form a semi-circular groove, in which the spermatic cord is received. The upper and inner walls of the canal are formed by the parieties of the abdomen.

This canal is about one and a half inches long, and terminates in the external abdominal ring. This ring, or rather this triangular opening, is formed by the tendon of the external oblique muscle, as it is separated into two distinct pillars or parts, one of which passes on to be inserted into the symphisis pubes, the other into the spine of the pubis. Escaped from this, the descent of the Hernia is direct along the line of the spermatic cord to the scrotum.

Thus we have seen what an Oblique Inguinal Hernia is; learned its course from the abdomen to the scrotum, and also the very important fact that the Inguinal canal has a lining of *serous membrane*, the prolongation of peritoneum, which the testicle had pushed before it from the margin of the internal ring.

It is an established fact in pathology that serous membranes, acted upon by proper applications, will throw out a coagulable lymph, which coming between two serous surfaces thus excited will cause them to adhere. Here, then, is the whole philosophy of cure by the Sherman method. The Hernia is completely reduced, and the support is applied while the patient is yet in a recumbent posture. If then the appliance is so adjusted *that it maintains its position over, not only the line of the external ring, but along the line of the canal*, irritation of the lining membrane of the canal must be the consequence, and the adhesion of the walls of the canal to the cord or the round ligament, as the case may be, and to each other, will follow as a natural consequence. This done, the hernial opening is hermetically sealed, so that the accident is not likely to recur.

To accomplish this end has been the aim of several truss inventors; as Hull, Semple, Stegral, Chase, Marsh, Seely, Schleiforth's Imitation-Marsh, Howe, the Elastic Truss Co., of New York City, the French Imitation Truss, and numerous others.

THE ELASTIC TRUSS CO.'S TRUSS.

We will examine the most popular of these before the bar of reason, setting down "naught in malice." This Company claim for their truss that it "is worn with perfect comfort, night and day, adapts itself to every motion of the body, retaining rupture under the hardest exercise, or severest strain until permanently cured."

This averment, I say, is a *specious delusion*, fatal to him who acts upon it; for just so sure as effect follows cause, will an *increase of suffering and displacement* follow the use of this instrument. What is its structure? A stout, elastic rubber band two inches in width, passing around the body, having at the end a *bulk of wood the size of a man's fist*, more or less, designed to cover the rupture, and prevent its passing out at the internal ring; another india rubber band, attached to the back of the first, passing down between the legs, and attached in front to the wooden bulk, *drawing it down heavily upon the pubic bone and spermatic cord, leaving unprotected the orifice through which the viscera descends.* Again the body being of an oval form, the effect of the unyielding, gripping band around it necessarily interrupts the general circulation by its great force around the hips—debilitates and wastes the muscular structure, destroying the symmetry of man's physical organization, and predisposing its victim to that most terrible affection, paralysis.

Look again at it closely and observe its mechanical action on the parts implicated in Hernia.

The bulk of wood, called a rupture pad, is very stout, and the pressure, by its attached band, necessarily very great.

Very great pressure, as we know, weakens and enlarges the
rings by a well established physiological law, thus *increasing
the difficulty instead of alleviating or curing it.* The more you
press the more you weaken and enlarge, and the more you
enlarge the greater the protrusion, until the sufferer, in an
agony of despair, exclaims, What shall I do for relief? Visits
Europe, perhaps, if able; exhausts the skill of its most
celebrated surgeons; tries on all the trusses to be found giving
the faintest hope of relief; but finding the *inevitable spring,*
common to all, returns, only to suffer, without the joy of
fulfilled hope. This is the picture presented by those who
have been duped into the use of these abominations, offered
with so much pomp and unconsciousness by the thousands
who seek the almighty dollar from the hands of the unfortunate, to whom they promise life, while they furnish that which
leads to greater suffering and irremediable injury. Independ-
ent of the discomfort they occasion, they injure the delicate
spermatic cord by the *extreme pressure* of the wooden chunk,
and thus crippling man in the most important and highly sensitive part of his organism — his procreative power — by
rendering him impotent, and powerless for social and domestic
happiness with woman.

Did I not truly say, this band was specious, but fatal and
dangerous. The mechanical action of its structure, warrants
the assertion, and the experience of the sufferers from Hernia
testifies to its fidelity. Again, look at the dreadful maladies
arising from displacement, all of which are aggravated by the
injudicious support of the truss. We have irritability of the
nervous system, gradually undermining the constitution;
Bright's disease of the kidney; irritable bladder, causing frequent desire to urinate; backache, general debility, loss of
energy and indifference to life.

· Have we not sustained our indictment against this pretentious charlatan — the Elastic Truss? To an intelligent public we leave the rendering of the verdict, confident that it will decide in favor of truth and humanity, both of which are against it.

SEELY'S HARD RUBBER TRUSS.

We come now to the consideration of the truss patented by Riggs and known as the Hard Rubber Truss of Seely. This is the old fashioned truss, simply covered with hard rubber. The material out of which it is made entitles it to our consideration only on the ground of mere *cleanliness*. In this respect alone it is unobjectionable. Outside of that it is deserving of all the condemnation bestowed upon all the other trusses, with the addition that the disagreeable and offensive odor arising from it when in contact with the heat of the body renders it very objectionable. As an article of manufacture it may be profitable to the unscrupulous vendors, who offer it to the public as though it really possessed novelty and merit. In order to meet the varied notions of those seeking relief, some are made in the half hoop form, with the tomato shape or oval rupture pad, and an oblique, flattened back pad. Others are made after the fashion of the old French truss, with a bulk at one end of a spring passing nearly round the body, and which is held in place by a strap completing the circle. The *first*, when adjusted, presses directly on the lower part of the spine and on the abdomen adjacent to where the rupture makes its appearance. Sometimes a strap is used for the purpose of retaining it in position. It is exceedingly liable to *displacement*, without the connecting strap, and even with it there is no security. The pads for the rupture are necessarily large, tending only to aggravate what they are recommended to relieve. When we look at it as a mechanical device, and contemplate its effects upon the body, we are led to denounce it and class it among the objectionable *palliatives*.

The *second*, or snake head form, has no more merit, and may be disposed of in the same category. If anything, it is more objectionable, the rupture compress being longer, coming more directly in contact with the spermatic cord and producing greater relaxation. Is it not strange that out of so large a proportion of the afflicted, these numerous appendages offered for the relief and cure of Rupture should be used without any consideration of the mischevious results necessarily following their adoption. It must be borne in mind that it is not the *material of which the hernial support is made, that is to meet the emergencies of the sufferer, so much as it is the* STRUCTURE *and adaptation of its parts to the relief and cure of Hernia.* The hard rubber finish is captivating to the eye and more durable and cleanly than leather, and in this respect *alone* does it possess merit over those after which it is modelled.

MARSH'S RADICAL CURE TRUSS.

This instrument, like the other pretentious claimants for popular favor, has its admirers — but it is destined to run its course at no distant day, and be buried in a common grave with its predecessors, by the general condemnation of the public. It has no merits of its own, for it is simply a combination of the concave pad of Hull, with the cone of ivory in the center in imitation of the leaden cone of Semple. Neither singly nor in combination do these elements answer the purpose of retaining the Rupture or effecting a cure. They have previously failed, after a faithful trial by the public, and the efforts to resuscitate them in a new form have alike proved impracticable.

Marsh claims that by this combination pad he holds the Rupture with greater security, and produces adhesive inflammation. This is theoretical rather than practical, for the severity of this half hoop spring, acting on this pad and the pad placed on the spine, for the purpose of producing this inflammation, has caused the most excruciating and horrible results. Now, if the purposes for which he claims this instrument is applied, could be accomplished, good might result from it. But, in view of the anatomical structure of the parts involved in Rupture, the suffering and injury produced are in direct antagonism to the sense and feelings of humanity. The *end* for which the truss was instituted is thus not only defeated, but the rupture greatly aggravated by the relaxation and enlargement of the opening for the descent of the gut.

Again, by bestowing exclusive attention on the Rupture pad, regardless of the tremendous lever power of the spring, these so-called truss-makers failed to perceive the displacing power they were unknowingly employing in constructing a vast lever whose fulcrum rests on the back pad, with the Rupture pad at the other end of the long arm — the spring. A blunder most fatal to their success, and which no mechanician of eminence would for a moment have committed. Fortunate for suffering humanity it is, that it, too, will ere long find its grave through the growing knowledge of the profession and the sufferers, and that no tears of regret will be dropped over its contemplated remains. At the same time indulging the pleasing expectation that the deluded public, who have suffered so much by its fatal conception, will exercise more vigilance in future over these pretended but ignorant guardians that aspire to treat their physical imperfections.

I might proceed further in the examination of the various trusses claiming popular support, such as Schliefarth's,

Howe's, Chase's, the French Imitation and the Boston Common Sense, but it would only be a needless repetition of what we have before said of those whose claims we have analyzed. The fatal defects are common to all. They bear not the gaze of scientific scrutiny, nor mechanical adaptation, and consequently are unworthy of the confidence of the public.

SHERMAN'S HERNIAL APPLIANCE AND CURATIVE COMPOUND

"One contrivance only, of which I have any knowledge," says a distinguished surgeon, "is not obnoxious to any of the above fatal objections." And as that one has about it so much to commend it to the profession, I may be allowed to give a more minute description of it. I allude to the Hernial Appliances of Dr. J. A. Sherman of New York City. The pads, both for the Rupture and the back, are fixed to a *composition* flexible band, which nearly encircles the body, so that when the instrument is adjusted, the back pads are applied on each side of the spinal column, *but there is no pressure whatever* on the column itself. From this point the spring passes around the body, having impressed upon it just vertical curve enough to lift it above the crest of the Ilium, the course of which it traverses until it reaches the anterior superior spinous process of that bone. Here a new curve is impressed upon it, which takes it directly over the course of Poupart's Ligament; it passes along the line of that ligament until it reaches a point over the external ring. From this point it rises gently over the symphysis pubis, and with similar curves to those already described, passes over the opposite side of the body to its place alongside of the spine, on the opposite side from that from which it started. Thus we see that the point directly over the external ring, where the rupture pad is of course placed, is the *stationary point* of the apparatus—THE FULCRUM OF THE LEVER—and not the long arm as in the other devices. Here is the application of a principle never before thought of, by which an immense advantage is gained over any of the many trusses in use. The Rupture support is oblong or ovoid, slightly convex, and ample in size, the greater diameter being placed on the line of the Inguinal canal, the larger end being applied to the external ring; so that the greater amount of pressure is applied there, which pressure is gradually diminished, until it passes over the internal ring.

This support is fixed to the main band by an ingenious contrivance, which enables the operator to apply the pressure, however, in any required direction and to maintain it there, or at will let it be diffused over the whole surface of the canal and rings, and yield to the motions of the body; in which case the greatest pressure is always on the most dependent portion of the Hernia.

Thus we have an appliance which answers every indication that can be desired by the surgeon. First, and most important, is the ease with which it is applied, and *its perfect freedom from the danger of displacement.* Next, the manner in which the rupture support is fastened to the main band, as already explained. Again, the fact that the pressure, being made on the rupture support at every mechanical advantage, need not be so great, and therefore can, with more certainty, be borne by the patient.

Another, and a very important consideration, is, that the material of which it is composed is not liable to rust by contact with the secretions of the body, nor to absorb its disagreeable effluvia — evils by no means to be disregarded by those seeking purity as well as efficiency.

The appliance is equally applicable to double or single Ruptures, as pads are fixed with equal facility on both sides.

Finally, to satisfy ourselves of the efficiency of these Appliances of Dr. J. A. Sherman, we have diligently inquired. of a great many who have used them, and have not yet received one iota of disapprobation. The uniform reply has been, We had been great sufferers for years; sought relief from the use of the Hard Rubber, Chase's, Marsh's, the Elastic Truss Co.'s and many others too, needless to mention. Now, by the use of Sherman's Appliance we feel a degree of comfort and safety never felt before. Many of those we examined we found cured and the Appliance laid aside. Others needed only time and prudence to accomplish the desired end.

Thus have we performed a pleasing and we hope profitable task, to the public, of laying before them information on a much needed topic. Amply repaid shall we be, if it shall prove to them a warning and a guide, and rescue them from the horrors of the surgeon's knife and the agonies of strangulation.

In conclusion, we may say that, being desirous of adding living facts in confirmation of the scientific views we have presented, of the vast superiority and efficiency of Dr. J. A. Sherman's Appliances for the relief and cure of Hernia, we made it our business to examine personally some who were present in the city of St. Louis, who were cured; and wrote to others at a distance, and now present the results of our investigations.

CASE NO. 1.

HENRY ROLOER, to be found at the Union Market, was ruptured on *both* sides *six* years ago—has been using trusses innumerable, with plasters. They only served to torment and aggravate his difficulty. In this condition he applied to Dr.

Sherman one year ago. *He is now perfectly cured. Of this we satisfied ourselves by a careful examination.* Any one desirous of similar convictions can call on Mr. Roloer, who will take pleasure in exhibiting his cure.

He expressed to the writer, as all others have done, great satisfaction with the treatment which enabled him to attend to his business, which is often laborious, with as much alacrity, from the first week, as though he had not been ruptured.

CASE NO. 2.

B. T. SMITH, connected with the well known house of Hickman & Wilson, Trunk Manufacturers, in St. Louis; fifteen years ruptured. In 1870 applied to Dr. J. A. Sherman, and was effectually cured in *eight months.* He is now perfectly cured, and has felt no symptoms of Rupture since. Mr. Smith made these *statements to us personally*, expressing at the same time the greatest confidence in the efficacy of Dr. Sherman's Appliances, and will cheerfully confirm all we have said.

CASE NO. 3.

REV. J. V. HIMES.—In a letter addressed to Mr. Himes asking for a statement of his present condition, he replies as follows:

"I use Dr. Sherman's Appliance, and have been *cured.* But I continue to wear it, because it is so comfortable and also *safe*, in case of strains or accident. I have had no pain or inconvenience since I put it on five years ago. It is the best instrument of which I have knowledge. I could not say more, as it would be a mere repetition.

Truly Yours,

J. V. HIMES.

Buchanan, Michigan, 1876."

CASE NO. 4.

CARSON CITY, NEV., March 8, 1876.

W. PRYOR, M. D., 1410 Olive St., St. Louis :

Dear Sir:—I am in receipt of your favor of the 28th ult., inquiring of me whether I was cured of Rupture by the Appliances of Dr. J. A. Sherman. My answer is, that I was.

While residing in Boston, in 1850, and in the express business, under the name of Fisk & Rice, office Court Square, I was ruptured on the right side. I purchased a truss of Dr. Phelps, on Tremont St., opposite the Tremont House, about 1863 or 1864, and since I resided here I became ruptured on the left side too, and was of course obliged to wear a double truss, and from this time on I suffered great agony from all the trusses I could find. I was on visit to Boston in the spring of 1871 and got a truss made by a brother of the Dr. Phelps who made my single one, but this, like all the double ones I had tried, soon gave me great pain. While subject to these torments I saw the advertisement of Dr. Sherman in the San Francisco papers ; opened a correspondence with him and then visited him, I think in April, 1873, and commenced his treatment. On the following March, about eleven months after, I abandoned the Appliances which he had given me, and have never used, or had occasion to use, either since. I am perfectly and fully cured of a rupture of over twenty years' standing.

I cannot too strongly commend to all afflicted in the same way, the use of Dr. Sherman's Appliances.

I have been a resident of this place for sixteen years, and there is not a State or United States official who does not know me.   Very truly,

H. F. RICE.

HOW THE CURE IS EFFECTED.

This topic is so important that, at the risk of some repetition we will mention the points under consideration, which are so necessary to be clearly understood, before intelligent action can be successfully undertaken.

Of the multitudes who visit Dr. Sherman's office, seeking relief and cure of Hernia, many express doubt, saying their physician said he knew of no cure ; that the ·trusses they had used had failed to give either comfort or retention, and were positive sources of annoyance and anxiety ; and that they had come to him to see what he could do, different from what they had received at the hand of others. To meet these expressions of doubt and distrust, and to inspire more confidence in the suffering public, we offer the following considerations :

So far as distinguished medical authority is of weight in the decision of this question, we have the positive assertion of Drs. Lawrence Richter, Sir Astley Cooper, Dorsey, and others, that rupture can be cured ; and this we present as a good offset to that incredulous part of the medical profession who have not made Hernia a specialty,

Recent investigations enable us to state that it can be cured, and we adduce in evidence of the assertion, the cures that Dr. Sherman is constantly effecting among his numerous patients. Gentlemen of standing in the city of St. Louis are willing to state, to any one desirous of making the inquiry, that Dr. Sherman has cured them. Gentlemen from abroad, whose word would be taken as truth, equally assure us that they have been cured by Sherman's method ; and if more evidence than this is demanded, you would not be convinced though one from the grave should arise and assert the fact.

That the members of the medical profession should be

incredulous as to the cure of Hernia, is not surprising, any more than the distrust manifested by the public at large. They have formed their opinions from the inefficiency of the miserable contrivances known as trusses, which, up to the present day, have been one continued series of mechanical blunders, and false assumptions of physiological principles. Nobody believed in the existence of the continent of America up to the time of Columbus, but he, by his genius and courage, made that disbelief a verity. Dr. Sherman stands in the same relation to the invention of a device for the cure of rupture that Columbus did to the New World. He has demonstrated its curability by producing appliances that act in harmony with known physiological laws. He has not merely the genius to devise, but the tact and executive ability to apply the offspring of that genius. His instrument does not press too much, and thereby relax, or too little, and thus not retain; but just enough to aid nature in her effort to support. And this is why Dr. Sherman's appliances accomplish a cure, and why they are so deservedly popular.

A committee of distinguished physicians, appointed to investigate the merits of blocks in the cure of Hernia, when asked, "What causes the occlusion of the hernial orifice in cases of radical cure? And, if local irritation and inflammation are not the sole cause of this occlusion, what is their value as auxiliaries in the treatment?" replied, that the serous membrane of the hernial sac never loses its peculiar tendency to adhesion under slight irritation, and that in most cases of the radical cure of hernia, the neck of the hernial sac is obliterated, either by adhesion or, more rarely, by absorption. Dr. Reynell Coates, in 1836, expressed the opinion, which was in harmony with the most distinguished surgeons of his day, that a hernial contrivance, to be successful, must possess perfect retentive power, and this, aided by moderate irritation, accomplish the result; and that none of the trusses of his period were worthy of confidence as a means of radical cure, because of their deficiency in that respect.

In this state of conflict among the M. D's. we leave the question of cure to be settled by experience. We are fully satisfied in our own minds, and have endeavored to present the grounds of that satisfaction, that any hernial appliance for its radical cure, which does not develop slight irritation of the peritoneum, will not only be inefficient, but fatally delusive, as a means of safety. Dr. Sherman's applications, which embrace both mechanical and medicinal agencies, fulfill the demands of the contestants perfectly, as has been certified to by Dr. Willard Parker and Dr. Carnochan, both eminent surgeons in New York, in the remarkable case of Jas. Corlew; and thus offer to the public that which both irritates the serous membrane and causes perfect retention, accompanied by a perfect skill in their adaptation of nearly forty years' successful practice in this one specialty of his life.

PART THIRD.

ORGANIC DISPLACEMENT.

ORGANIC DISPLACEMENT.

PROLAPSUS UTERI.

In our treatise on Hygiene, we spoke of the division of the body into organs, and considered the most important in detail, with the exception of the womb, which is peculiar to woman. We now propose to examine this feminine peculiarity, setting forth its *grandeur and dignity*, with the *abuses* to which it is subject in our so-called age of civilization ; the *causes* of those abuses, and the *remedy* for their removal.

Every object in this broad universe of facts and phenomena has a *definite constitution*, and stands in *fixed relations to every other object.* This constitution and its varied relations, indicate the purpose for which they were made, and are governed by *laws*, in virtue of which these objects act and react upon each and every other.

These laws are *permanent, uniform and universal*, and are called Natural Laws. *They are the stereotyped will of Omnipotence*, graven into the *nature of things*, and can no more be *violated without suffering than we can put our finger in the fire and escape pain. Nor can we any more obey them and not reap enjoyment*, as the reward of that obedience, *than we can eat delicious fruit with a healthy palate*, and not feel gustatory delight.

Man, in virtue of his constitution, has a fixed purpose to subserve. Woman, likewise, in virtue of her constitution, has a fixed end to serve.

It is clearly written in the mental and physical organization of man, that the grand *primal* end of his being is to *originate life*—all else is incidental;—and this he does through the instrumentality of *semen*, a fluid generated in a pair of organs called the Testes. Here, in this great life workshop, are created those germs called "spermatozoa, each of which is a nucleus of a human being—with its spinal column, and every other part, whether mental or physical, peertaining to a developed male and female."

What a grand function, this origination of life! What momentous consequences hinge on its development! How few there are who know of its grandeur; and of those that do know, how few there are that appreciate their responsibility, in regard to the gift. Rightly considered, 'tis the grandest act in all nature, for from that act come all of our earthly joys, our sorrows, and our sufferings—joys, when life enters on the scene of conflict, prepared to cope with its grave responsibilities and sorrow and sufferings innumerable, if, with a feeble body and diminutive brain, one begins the voyage on life's troubled sea. Such has been the picture presented by the entire race, from the first transgression down through all the ages to the present. Woman, in virtue of her specific physical organization, the *womb* and its appendages, is destined to *receive, and feed and develope these life germs.*

Maternity, says Mr. Fowler, is the door through which all that lives enters upon its terrestrial existence. *Womb-man* is is one of the most expressive of all Saxon words; its first syllable designating that fountain from which gush forth whatever qualities appertain to the entire female sex as such, Every iota of female beauty comes from it. When it is impaired, all her beauties of form, complexion, face, bust, limbs, pelvis, decline with it. She justly sets all the world by her personal charms, graces and accomplishments. Then let her realize that this is their only anatomical source. If there is any entity, the most sacred and holy on earth, *womb* is that entity. Sacred to woman, as the source and centre of everything feminine and holy; to man, as that in which originates all that is loving and lovable, in the female sex in general, and in his own wife and dear little ones in particular.

Here is a picture of its form, which admirably adapts it to the purpose of its creation;

FIG. 1.

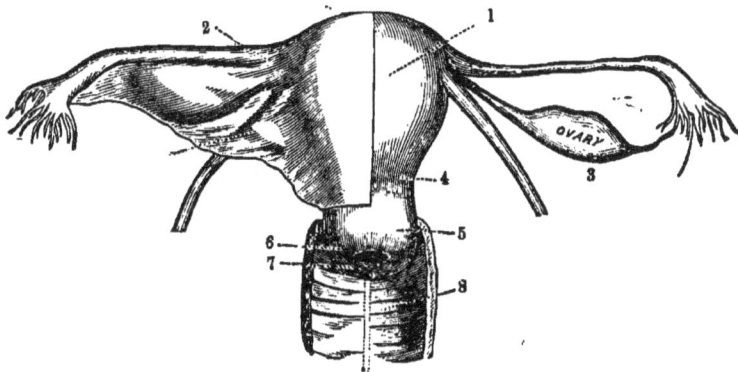

1 Fundus.	5 Vagina.
2 Fallopian Tubes.	6 Anterior Lip.
3 Ovary.	7 Posterior Lip.
4 Broad Ligament.	8 Interior of Vagina.

By its position it it perfectly adapted to the growth of the life germ, being protected by surrounding viscera and bones' and kept at a uniform temperature. It has three coats—an external serous; a middle one, composed of layers of muscles, which enter most largely into its bulk, and an internal one of mucous. Its appendages are the *ovaries*, where the life germs, called eggs, are manufactured; the *fallopian tubes*, which convey the eggs from the ovaries to the womb; and the *muscular ligaments*, which sustain the womb in its place.

THE FEMALE PELVIS

AND

ITS SURROUNDING ORGANS.

FIG. 2.

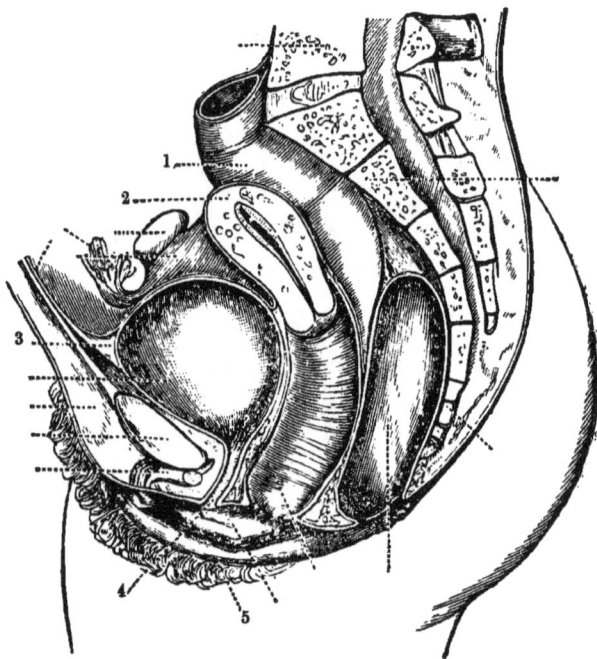

Fig. 1 The Rectum.
" 2 The Uterus.
" 3 The Bladder.
" 4 The Urethra.
" 5 The Vagina.

FIG. 3.

PERFECT, HEALTHY, OR
NATURAL FORM.

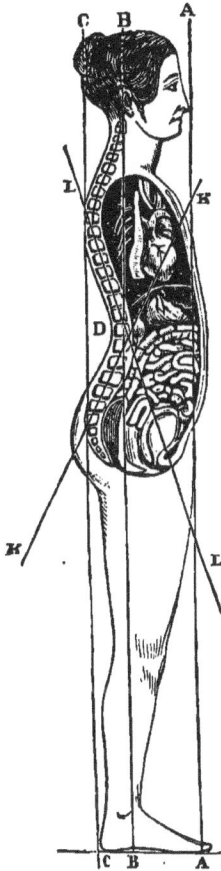

FIG. 4.

DISTORTED, UNHEALTHY,
OR UNNATURAL FORM.

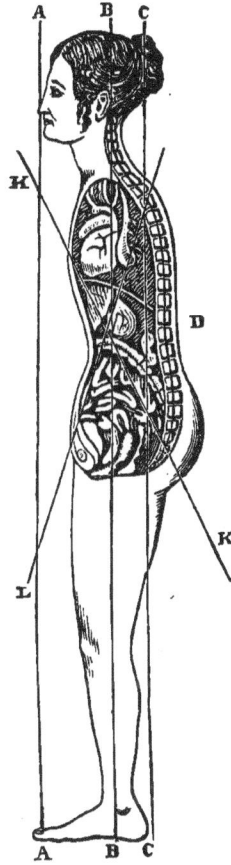

The above cuts are designed to show the position of the viscera in the great cavities of the body—the chest and abdomen. In the perfect, healthy or natural form, the chest and spine have the beautiful bow-like curve. The heart (h) rests on the diaphragm (d), which is also curved upwards. The abdomen is prominent, and the stomach (s) is supported by the intestines (c).

In the *distorted, unhealthy or unnatural form* we see the beautiful curves of the chest and spine wanting, the heart unsupported by the diaphragm, which is nearly horizontal, and the stomach unsupported by the intestines ; all of which *necessarily causes the viscera to sink downwards.*

The uterus is situated in that region of the body known as the pelvis. It it placed *between* the bladder in front and the rectum behind, above ; and, pressing upon it, are the viscera of the two great regions, the abdominal and thoracic.

The uterus, we have seen, adapts woman to her specific function—that of sustaining and developing the life germs. It is the chamber and home in which, for a period of nine months, this developing process is carried forward. At the expiration of this time, the young embryo is brought into the world. *Previous to conception* the woman manufactures *more* blood than the wants of her own body demand. This *surplus blood* passes off from her womb, if she be healthy, with the natural degree of regularity—once every month. These periods are designated as her "courses, menses, catamenia," which "*usher in*" and "*close out*" womanhood.

So long as this flow is regular, of the proper color and consistency, and in due proportion, the female system rejoices in its healthful vigor ; but the very moment derangement ensues, then comes the headache, chilliness, numbness, cough ; and, if the suppression continues, the whole health of the female suffers, with the almost *certainty* of a ruined constitution, *unless* a restoration is effected. Hence the great importance of this function, and the care necessary to be bestowed on the health of the body, in reference to its regular performance.

As *soon as conception* has taken place, this *surplus food ceases to flow*, for it is *now* needed by the young embryo during its ante-natal state.

OF THE DISEASES TO WHICH THE WOMB IS SUBJECT.

We may name among the many diseases to which this most important organ is subject, all of which are characterized more or less by hyperaemia, or congestion, and terminating, if neglected, in inflammation and ulceration ; as

Cancer,

Fibroid Tumors,

Growths of various kinds,

Displacements.

In 1200 cases examined by Dr. Hewitt, of London, there were 714 with uterine disease, and of those 81 were cases of Prolapsus and 54 of Cancer.

It is our intention only to examine that distressing and most prevalent affection, Prolapus of the Womb, considering its symptoms, causes and cure.

Prolapsus means falling of that organ *within* the vagina, or its appearance external to the vagina. *Its attendant symptoms and disorders* are languor, lassitude, weakness, fatigue on walking a short distance, indisposition to rise in the morning, weight in the pelvis, irritation of the rectum, and pain in the back and loins ; pulling at the center of the chest, and a drawing downward at the pit of the stomach, dyspepsia and biliary troubles ; quick and short breathing, "palpitations, flutterings and sinking feelings."

"But," says Miss Catharine Beecher, "the '*lower intestines*' are the greatest sufferers from this dreadful abuse of nature, having the weight of all the unsupported organs above pressing them into unnatural and distorted positions. The passage of the food is interrupted, and inflammations, indurations and constipation are the frequent result, and one in which both sexes are equal sufferers. Dreadful *ulcers* and cancers may be traced, in some instances, to this cause."

Again, this distortion brings on woman peculiar distresses. The pressure of the whole superincumbent mass on the pelvic organs induces sufferings proportioned in acuteness to the extreme delicacy and sensitiveness of the parts thus crushed. And the intimate connection of these organs with the brain and whole nervous system renders injuries thus inflicted the causes of the *most extreme anguish both of body and mind. This evil is becoming so common, not only among married women, but among young girls, as is a just cause for universal alarm.*

CAUSES OF THIS PROLAPSUS.

These may be referred to the customs of civilized life, which have depreciated her powers and caused that general debility and relaxation of the muscular system that lies at the foundation of the displacement under consideration; as

1. Want of air and exercise.
2. Excess of the nervous system.
3. Improprieties in dress.
4. Imprudence during menstuation.
5. Imprudence after parturition.
6. Prevention of conception and production of abortion.
7. Marriage with existing uterine diseases.

Firstly—WANT OF AIR AND EXERCISE.—"This topic takes the lead of all others in importance and difficulty. The fact that the Greeks lived most of the year out-doors; and that in their *houses they never breathed any but pure air*, gave them an advantage, in developing the *beauty, strength and health* of their children, which it is not so easy to secure with our cli-

mate and habits.   And the steady and equable climate of the
old countries, which has led their inhabitants to out-door life,
and thus secured vigorous constitutions, gives them also a
great advantage over us.'

We have shown, in the treatise on Hygeine, that every pair
of lungs vitiates a hogshead of air every hour by withdrawing
half of *its oxygen, the vital principle, and replacing it with
carbonic acid gas, a poisonous principle.*

A few years ago a most interesting report was presented to
Congress on " Warming and Ventilating the Capital." This
report was prepared by Professor Henry T. U. Walter, and
Dr. Wetherill, and was accompanied by a large collection of
carefully arranged tables of the " Analysis of Air." These
tables were prepared from actual experiments, made by scientific men of various nations, each giving the result of his own
analysis of air, taken from great elevations, on mountains and
in balloons, from mid ocean and from the coast, from valleys,
from the center of the continent, from populous cities, from
the open country, in winter and in summer, during the day
and at night, and also the comparative analysis of the air out
of doors and in occupied buildings.   ONE OF THE MOST FORCI-
BLE FACTS BROUGHT TO VIEW IN THESE TABLES IS THE UNIFORM
PURITY OF THE EXTERNAL AIR ALL OVER THE GLOBE, EVEN IN
DENSELY POPULATED CITIES.

Here we have, at last, the secret unveiled of the ill health
of our American women.   We *must look to the illy-ventilated
condition of our houses, in which the American woman is
doomed, by our false views of civilization,* for their growing
deteriorated health, and especially that form of it we are now
considering—prolapsus of the womb.   This hogshead of pure
air must be had every hour of our lives to cleanse this Augean
stable of our bodies, and it is by means of the oxygen gas it

contains that this grand work of health, strength and vitality is mainly accomplished. The homes do not furnish it; and woman, who is doomed to remain so long in these homes, suffers through her entire organism.

Every householder should be sure that every member of his family breathes pure air, not only all day but all night, by this simple arrangement: In every room of his house let at least one window be let down at the top two inches, and one door have an opening of two inches over the top. Let this be done in such a way that no one *can* alter it. A house thus arranged will require more fuel to warm it, but the additional expense of this will not be a tenth part of that which would result from the loss of labor and health consequent on the debility and disease always resulting, more or less, from the habitual inhalation of impure air.

As to unhealthful *miasmata* in the *night air*, nothing can be worse than the exhalations of decaying bodies as sent forth from the lungs and skin of sleepers. It is precisely the same evil as is found in proximity to grave-yards and decaying carrion. Those who have entered the pent-up sleeping-rooms of persons who do not wash their skins or breathe a pure air, very well understand the close resemblance.

The school-houses, halls in public buildings, churches and railroad cars, in short, every apartment in which a human lung is doomed to breathe, *must* have this hogshead of pure air provided every hour of its breathing life, or debility, sickness, disease, impaired constitutions and premature death will inevitably follow.

To the members of the medical profession, who are considered the guardians of the public health, we especially commend a more attentive consideration of the relations of oxygen gas to health and disease, and that they will *insist* upon a more faithful application of these relations in the public and private dwellings of the land. It is through such practical enforcement of vital laws that we can alone hope for an increased elevation of our race in the physical and moral health essential to the perpetuation of our institutions.

Next to pure air, suitable exercise and amusements are essential to the health of woman. Their absence now causes the malady under consideration; for, be it remembered, *debility* is the main cause of its existence.

Under the head of Muscular Activity, we gave a picture of the effect of *inaction* on all the organs of the body. We then showed how *action* put in *motion* the *blood*; how the blood was purified by the increased respiration thus promoted; and how this purified blood was sent dancing along the circulation to every part of the body, giving increased life, energy and health to every organ. *The blood, we repeat, is the life of all flesh,* and without exercise, there can be no distribution of the life current, and no growth, health and vigor of any part. Let woman provide herself with suitable *out-door* employment, or, if in,door, with *pure air*, and let her in the prosecution of it, call her muscular system into *action;* and let all this be done cheerfully, so as to send down from her brain, through her nerve-tubes, this precious life-giving nerve force, and we will be responsible for the beauty, health and symmetry of her body, and no longer see her the victim of pale cheeks, distorted form, wasting consumption and prolapsed womb, &c., &c.,

A distinguished lady, Miss Catharine Beecher, has most forcibly said, "Not in the wide circuit of our nation is an institution, where *one* teacher is sustained, whose official duty it is to secure the health and perfect development of that wonderful and curious organism on which the mind is so dependent. The students in our colleges, and other institutions of learning, should be required to breathe pure air ; to exercise their muscles appropriately and sufficiently ; to retire as well as rise at proper hours ; to take care of the skin ; and to avoid the use of stimulating herbs and drinks ; and the same watch and care should enforce these duties as are now devoted to training the intellect. An endowment should be provided to sustain well-qualified men and women, whose official duty it should be to give instructions, and exercise the supervision that should secure so important a result."

Secondly — EXCESS OF THE NERVOUS SYSTEM.—We have seen, in our article on the brain and nervous system, that from these there originates a force, or energy, which controls the entire body ; and that upon its quality and quantity, and proper distribution to all the subordinate organs, depend the health and vigor of every one.

Proportionate action is the law; all *excess* is injurious, and tends to undermine the constitution. Exercise the brain too much in study, or too much care or anxiety, or grief, and the *balance* of nerve force is broken. The brain suffers by excess, and stomach, liver and bowels, &c., by diminution, and we thus have dyspepsia, and biliary diseases, and great debility. Violate any law of health, and you tax the brain by *unnatural and excessive action to repair the injury.* If the stomach is overtaxed by too much food, or food that is inappropriate, the brain is taxed unnaturally, to send down to the suffering organ more nerve power to relieve the stomach of its burden.

Be it remembered, that every drain upon the nervous system, is a slow but sure undermining of the constitution, which shortens life and renders the body more subject to diseases of every kind.

The constitution of woman is naturally more delicate than that of man. Her nervous system is more sensitive, and demands more care for its healthful performance. But society, ignoring her peculiarities of organization, has failed to give her such an education as will strengthen her mind and body, and protect her from the physical dangers to which she is liable. Fashion loads her with her pernicious forms of dressing; late hours rob her brain of rest; ill-ventilated sleeping-apartments and households poison her body with impure air; and an aimless, unoccupied, unexercised daily life, takes from her that wholesome mental stimulus that gives tone and vigor to her organs.

She is therefore, unfortunately, illy prepared to resist diseasing causes; and when assailed by cold, or taxed by the cares and consequences of marriage, she falls a victim to displacements, which embitter her life.

Thirdly. IMPROPRIETIES IN DRESS.— This has been a fruitful theme, and justly so, for denunciation on the part of female reformers. Unfortunately for woman, science has had but little to do with providing her suitable garments. They have been devised by an ignorant fashion, whose sole aim has been to please the vulgar, uneducated eye, regardless of the more important and lasting effects upon her health, beauty happiness, and *offspring*.

She has therefore been, through ages, a constant victim to whims and caprices that have always led her into physical transgressions, causing the most distressing maladies. Among these, the most conspicuous is prolapsus.

When we regard the position of the womb, 'tis not difficult to perceive how this has happened. It is in the centre of the pelvic region, between the bladder and rectum; above are the intestines, liver, stomach and spleen, and higher up, the lungs and the heart. In health, these are nicely packed and adjusted, so as to mutually aid and support each other, with the help of muscular bands or ligaments. Let, however, undue pressure from above downwards be used; or let debility of the muscular system arise; or cause too much heat around the hips and abdomen, and the womb suffers instantly by the superincumbent weight, and undue warmth, and in process of time falls down, with its neck in the vagina. This undue pressure and warmth is effected by the modern style of fashionable dressing. Instead of being suspended from the shoulder, it is suspended from the hips. Instead of being diffused over the body, it is crowded too much upon the back and abdomen, and instead of giving facility of motion to the muscles and chest, it oppresses their healthful movements by undue constriction. Hence arise weak stomach, feeble lung power, and a muscular system too debilitated to allow of the necessary exercise to develop a healthy, symmetrical body.

Should it be wondered at, then, that this hydra-headed monster—Fashion in dress—should mar our present civilization, by ruining its mother, woman, in that specific life-giving organ, the womb? Surely no one can question its power after a consideration of its effects.

Fourth Cause—IMPRUDENCE DURING MENSTRUATION.—The importance and design of this function have been dwelt upon, and the reader, if not posted in regard to them, will please re-read them before entering on the examination of this topic.

Having carefully considered what menstruation is, and its uses in the economy of woman, he is prepared to understand *how* its abuse produces the physical displacement under consideration. We will explain.

The uterus being designed as the home of the embryo child, where it is warmed and fed, and where it grows and is developed, during a period of nine months, it was necessary in its anatomical structure to endow it highly with blood vessels and nerves, to accomplish this grand end. During the work of menstruation, it is charged with more blood than is needed in its ordinary state, and is therefore heavier, and more predisposed to congestion and inflammation. Let any cause, as cold for example, now be taken which prevents this flow, thus surcharging the blood with an unnatural excesss of morbid matter, and we all know the fearful wreck it tends to produce—and does produce if neglected—in the whole female economy. Ill health comes on and general relaxation follows, and with this relaxation comes the prolapsus. Hence the great importance of attention to the monthly periods — by watching over the general health.

Fifthly—IMPRUDENCE AFTER PARTURITION.— The entire physical system has been taxed to develop the embryo child. If the laws of health have been duly observed by the mother, she is prepared to stand the drain upon her physical stamina by the birth of her babe. But, as is most usually the case, with American women, the period of pregnancy is one of extraordinary neglect of nearly every physiological requirement. But little exercise has been taken ; little attention is paid to a regulated diet ; the nervous system never watched over ; the skin neglected ; and last, but not least, the dress, especially in fashionable life, presses so unequally upon delicate organs as to produce the evil so much to be guarded against, a downward pressure upon the womb and its neighboring viscera.

Now let a woman *rise* too soon after the birth of her babe, with her muscular system weakened and relaxed by the stretch just taken from its fibers, and we see at once the danger of displacement

Again, it is known by the enlightened physician that at least two months are required for a natural restoration of the womb to be accomplished. If COITION is indulged in *before* this healthy restoration is effected, irritation of the womb is established, and this leads to congestion and this to prolapsus.

Let the husband who values the sacred office of his wife's womb watch his imperious sexual instinct, and see that it, by its illy-regulated action, does not defeat his own happiness by ruining the health of his dear wife.

Sixthly—PREVENTION OF CONCEPTION AND PRODUCTION OF ABORTION.—It is not difficult to conceive *how* this cause operates in bringing about the terrible evil of prolapsus. For when it is known that acts of mechanical violence to the womb are the usual means to accomplish this wicked end, we see at once and clearly the fatal mischief wrought upon the womb, and through it upon the whole constitution, thus exhausting the vital howers and prostrating its strength. The support which holds up the womb is consequently withdrawn, and the inevitable prolapse takes place.

Seventhly—MARRIAGE WITH EXISTING UTERINE DISEASES. —It is sad, but nevertheless 'tis true, and should therefore be told, because if withheld great mischief will result to woman, whom, we have seen, is a co-operator with man in transmitting life and health and energy to her offspring; that female life is far from being in harmony with the grand laws for the development of her physical and mental powers. Woman is therefore weak, and necessarily subject to disease. That her diseases are multiplying is a matter of fact, and revealed through the medical profession.

Marriage, to be productive of its high purpose, the bringing into life a sound, well-formed, healthy human being, should be undertaken *only* when both parties are in the highest state of health. Any imperfection, whether of mind or body, is a legitimate ground for its non-performance. Should it therefore be undertaken with any form of disease, and especially with those known as leucorrhœa and amenorrhœa, the high transgression is visited with increased suffering, and prolapsus is one of the probable consequences of the violated law. From what we have written, this result is a clear deduction from our premises, and no sensible mind will fail to perceive the connection between them.

RESUME.

To re-impress what we have written upon a subject of so much importance, and of such increasing frequency as prolapsus of the womb, we will re-state the positions we have taken briefly, and the grounds on which we have based them. We have seen that the uterus and its attending ovaries are the characteristics of the female organism; that they are designed for the development and growth of a human being, and that they are placed in the pelvic cavity *between* the bladder and rectum, and suspended by ligaments attached to the surrounding parts, and pressed upon by the superincumbent viscera of the two great cavities of the abdomen and chest, and bound in by the abdominal and dorsal muscles; that in a perfectly healthy subject these viscera are so admirably packed and supported that a case of prolapsus of the womb is impossible. Consequently, whenever a case of falling of the womb does occur we may trace its cause to the conduct of life in reference to one or more of the following heads:

1. Abuses of air and exercise.
2. Abuses of the nervous system.
3. Improprieties in dress.
4. Imprudence during menstruation.

5. Imprudence after parturition.

6. Prevention of conception and production of abortion.

7. Marriage with existing uterine diseases.

No woman who observes in the conduct of her life correct physiological habits in reference to the above topics will ever be afflicted with prolapsus. This is the testimony of the most enlightened authors in England, France, Germany and America. And we readily admit the soundness of this opinion, because when we reflect that the Author of our being is too good and wise to have made us for disease and premature death, but rather for health, happiness and long life, (and this all attain who understand and obey his laws), no other conclusion would satisfy the requirements of the understanding but to turn to our habits and conduct in life as the main-spring of our misfortunes.

THE TREATMENT OF PROLAPSUS.

After a due consideration of the nature, symptoms and causes of this malady, we now come to an exposition of the various plans that have been proposed, from time to time, by the profession, for its relief and cure, showing their inefficiency and consequently failure to answer the purpose for which support should be applied. (We will then offer, for popular acceptance, the Appliance of Dr. J. A. Sherman, of New York, with the arguments in favor of its adoption.)

In considering the means for the relief and cure of the displacement under consideration, there are four indications to be fulfilled, says Dr. Hodge, of the University of Pennsylvania:

1st. To remove or palliate any existing cause.

2d. To replace the organ.

3d. To maintain at all times in *situ naturali* the womb, and to allow of its natural motions.

4. To strengthen the natural supports of the womb.

The first has been fully treated of, and must be carefully attended to by the patient. To the third we mainly direct our attention, as this is the mechanical indication to be fulfilled, and offers the greatest difficulty in the way of cure. This problem is *the one* for solution — for after all the attention and ingenuity, and even science, which have been directed to this point, in times past and present, no suggestion has received the general sanction of the profession. Innumerable as have been these suggestions, each has but a limited number of supporters. Many physicians have avoided such cases entirely, surrendering them to every variety of empirical experiments, so that women remain too often wretched sufferers, spending days, months and years a prey to diseases which disturb every corporeal function, as well as the whole intellectual and spiritual being. There has been no lack of ingenuity, and no want of experiments, with more or less success, but these efforts have not been generally well directed. The proper position of the organ in health, its means of support, and the suitable scientific indications to be kept in view for the relief of displacements, have not been sufficiently developed. Mechanical arrangements, it must be remembered, operate on vital tissues endowed with great sensibility, and if they *exalt* this, in connection with the superincumbent viscera pressing on the womb, they defeat the end for which they are used.

External bandages, such as braces, abdominal or utero-abdominal supporters, abdominal corsets, spinal supporters, have all been used from a period of time far back in the past. The weakness in the loins, emptiness or vacuity in the abdomen, the bearing down, the sensation of openness as if everything would escape, have prompted their use. But an enlarged experience does not justify the approval of patients.

They are but temporary palliatives, and have been found liable to the following objections : They have to be removed and replaced incessantly ; exert too much *friction* on delicate parts ; absorb the perspiration and other secretions, which requires the frequent renewal of them, all of which effects are greatly aggravated in warm weather, especially in corpulent females and laboring women. They do not raise the uterus above the pubis, their pressure being counteracted by innumerable causes, nor do they restore the organ to its proper place, but rather aggravate the pressure on the womb and increase its deviation.

Dr. Hodge says that the idea of replacing or supporting a displaced uterus, by pressure above the pubis, must be regarded as an absurdity in physics and deceptive in therapeutics.

The fourth indication is attempted to be solved by the use of Pessaries, Internal Supporters, or Intra-Vaginal Supports.

The most opposite opinions prevail among the profession in regard to their value. Dr. Hodge thinks if they are constructed of suitable material, and of the proper size and form, they will, if properly applied, remove all disagreeable sensations ; while Dr. T. Gaillord Thomas, of New York, an authority equally worthy of respect, deems them mere palliatives, and not in the least curative. For, " while they sustain the prolapsed organ temporarily, by their *bulk*, they prevent the vagina from contracting, and in time becoming capable of resuming its duty." The Pessary, he says, should support the uterus and not distend the vagina.

Many consider them painful, irritating and inflaming, causing ulcers of the *os uteri* and vagina.

In the preceding dissertation we have taken a systematic review of the displacement under consideration, and directed popular attention to the *causes* in operation, producing this increasing and most distressing *effect*. To remove an effect we must remove the cause from which that effect flows. Any treatment which aims at the accomplishment of a *permanent* good by any other proceeding is simply empirical, and fails in the attainment of its end. That this chronic affection, among a large class of others, has steadily resisted the treatment of the routine practitioners of the healing art, is very obvious to the ordinary observer of facts.

This of itself should have directed the attention of the thinking mind to the *reason* of so uniform a result in both the old and the new world. As Nature is uniform in her operations, we would expect to find on inquiry that *her plan* for bringing about results had in this case, as in all others where good is not attained, been little attended to, and a mischievous and disturbing one, of man's device, substituted in its place; thus perpetuating the evil, instead of removing its existence, as a standing reproach to her wisdom and benevolence.

We shall accordingly find that the seven active causes in operation in our enfeebling civilization, that we have endeavored to bring prominently before the suffering, have been too little attended to, and that the present plans of mechanical and medicinal aid have not operated in harmony with her physiological laws of cure. Hence, the stubborn resistance of this enemy of woman's domestic bliss and marrer of man's enjoyment.

We recommend, as a substitute for the pessary and brace (now so generally used), and all the irritating agencies of the schools, the following ingenious contrivance of Dr. J. A. Sherman, of New York City: It is constructed of a flexible band, passing nearly round the body, with a support attached at the front, adjustable by a spring which adapts itself to the abdomen, and gives a *firm lifting support, by which the womb is kept in its natural position.* There are pads placed on each end of the band, resting on the back, holding the support in its proper position. This appliance may be regulated to suit the degree of lifting support, to meet the requirements of the case. *It works upon the principle of applying the two hands to the lower part of the abdomen, and lifting upwards and ackwards* the womb, thus *assisting the relaxed abdominal muscles, and thereby affording injured nature a chance to recover its original vigor.*

The appliance is remarkable for its simplicity and efficiency, being adapted and removed with the greatest facility. This, in conjunction with medicinal treatment, without taxing the body with internal foreign devices that all the past has utterly repudiated, completely restores the patient, when taken in time,—otherwise, we can only promise, a great helper and palliative.

PART FOURTH.

—

HYGIENE.

HYGIENE.

That department of medicine which relates to the preservation of health is styled Hygiene. It is usually treated of in our works on Physiology under the *Laws of Health*. Health is not appreciated as its importance demands. Few deny its value—scarcely any, however, observe its rules systematically—nearly all come short of their requirements. Few are consequently robust, few are long-lived, few happy, and still fewer but what have aches and pains from the cradle to the grave. Health is indispensable to every form of enjoyment, every species of usefulness. The sailor, the soldier, the plowboy, the lawyer, merchant, statesman, doctor, and the mother, son and daughters, *all* demand, for the *efficient* performance of their various and responsible duties, health. Did the doctors *practically* observe and enforce the requirements of the *laws* of health; did schools, colleges and churches teach their absolute and unconditional observance as *antecedent* conditions to the sound performance of all physical and mental pursuits, the practice of virtue, and the attainment of long life and success; we should then witness a more promising aspect of society than at present is exhibited to the gaze of the philosopher. *Now* the future of the race is by no means encouraging. Weakness and debility, increasing from generation to generation, instead of strength and vigor, are the rule. Short lives instead of long lives; feeble-minded men and women instead of Bacons and Newtons and Fultons and Arkwrights; pigmies in body, instead of Anaks; in short, a physically and mentally deteriorating race of men and women, simply because the laws of growth and development of man are not systematically taught and obeyed by the nations styling themselves civilized.

Nature, however, keeps a faithful account with her children. Every transgression of her laws is recorded faithfully in the Book of Life. Regular debits and credits are made, and though she may not *seem* to demand a settlement for her violated commands or laws, there *is* a punishment for every command violated, and a reward for every one obeyed. Because the punishment is at first slight, it is too frequently overlooked ; but when these slight transgressions accumulate from day to day and month to month they form an aggregate of violated law that prostrates the subject and threatens life, and then, when it is too late, rouses attention to the perilous brink on which we are tottering. "An ounce of prevention is worth a pound of cure," is a wise adage ; prevention is better than cure, is another. Let us then betimes teach our children a knowledge of their bodies, a knowledge of their minds, and of that beautiful universe in which their bodies and minds are placed, and of those unchangeable relations between these minds and this universe of matter, out of which comes all of our health, all of our life, and all of our earthly joy.

OUR BODY.

The human body is divided by anatomists into bones, muscles, nerves, lungs, heart, liver, arteries and veins, stomach, spleen, intestines, bladder, kidneys and skin ; and when woman is the subject of inquiry, we have superadded the womb. This classification is sufficient for practical purposes, and as our object is utility, we propose to adopt it as the basis for our dissertation on Hygiene. We will now consider each anatomical element separately, and examine the laws for the healthy growth and development of each.

First—HYGIENE OF THE BONES.—The *hardness, strength* and *insensibility* which characterize the bones in a healthy subject, fit them in a remarkable degree as a basis of support to the softer and more active tissues of the body. It is by means of these qualities that the human frame is enabled to combine symmetry of form with freedom of motion and security to life.

The bones of the skull and the socket for the eye, serve exclusively the purpose of protection to the highly delicate and important organs within them. But the far greater number are designed for voluntary motion, and serve incidentally the purpose of protection. As a machine designed to execute the greatest variety of movements, we would expect the human frame to consist of component parts the most numerous and varied. Accordingly no piece of machinery is so wonderful in its combinations ; no piece of art can be compared with it for nicety in its evolutions, and all executed simply by muscular force acting on the bones and altering their relative positions. The position which man occupies in relation to the material universe, required that he should execute the greatest variety of movements, and to accomplish this end, it was necessary to give him an osseous structure, consisting of a great number of parts, and so connected with each other by articulations as to produce exactly the kind of motion the animal required from it. If the frame-work of our bodies consisted of one entire piece, we should be incapable not only of motion, but every shock we sustained would have passed undiminished to the whole frame. *Now*—by its division into parts, and by placing soft, elastic cartilages between them, and united by ligaments,—free and extensive motion is secured, and the force of every shock is deadened and diffused over the entire body, just as the interposition of springs in a carriage produces ease of motion while passing over a rough road. Thus, by this admirable arrangement is safety secured to our vital organs within the frame-work, and thus are our ideas of the wisdom and goodness of our Creator heightened and our veneration for his plans deepened. The entire collection of all our bones, united together in their natural order of arrangement is called the skeleton. They number in all two hundred or more—each is distinct from, but closely connected with, the rest, and of a shape, size and construction in exact harmony with the kind and extent of motion it is destined to perform. Anatomists recognize three great divisions of the skeleton—the head, trunk and extremities. The first is well known; the second includes the two great cavities, the thorax or chest, and the abdomen or belly; and the third comprises the arms and legs, or upper and lower extremities. Bones are composed of two kinds of substances, animal and earthy—to the first belong the attributes of life and growth; to the second the hardness and power of resistance they possess in a healthy state. In 100 parts, according to the analysis of Bergelius, 32.17 per cent. is animal matter, and consists of albumen, gelatine, cellular membrane, blood-vessels, nerves and absorbents. The remaining 67 per cent. is earthy, consisting of 52 parts of phosphate and 11 parts carbonate or lime. These two constituents, however, vary at different periods of life. In infancy the animal portion is in excess, and gives rise to that comparatively soft, yielding and elastic character, which distinguishes the bones of young children. In middle life the animal and earthy are nearly equally balanced, while in old age the earthy is largely in excess, and gives rise to that dry, brittle and comparatively lifeless condition that characterizes the bones of the aged. The ends accomplished by this arrangement are wise. In early youth, when much strength is not needed, as the body is not taxed with severe efforts, but when great growth of bone is demanded, to secure the development of the frame, the animal, or living part, is in excess. In middle life, when growth is finished and resistance is at its height, and nutrition is required to repair waste only, more of the solid or earthy is demanded, and less of the animal or vital constituent. When old age comes on and the wants of the body are reversed, and waste now exceeds nutrition, to put the frame in harmony with the shrunk muscles and feebler life-power, more of animal or life matter is taken away by the absorbents, leaving the earthy, which demands less support from within. Hence, the brittle and compact hardness of the bones, and their feeble power to unite when fractured, at an advanced age. At birth many of the bones are of a cartilaginous structure ; as age advances the cartilage is gradually removed, and its place supplied with cellular membrane, in the interstices of which the earthy matter or lime is deposited,—the two forming bone.

To the life or growth of every living structure two processes are essential—*waste* and *renovation*. To carry forward these, all parts of the body have *arteries*, which carry the red or nutritive blood ; *exhalants*, which deposit the nerve matter ; *veins*, which carry the blood back to the heart ; *absorbent vessels*, which take up the waste matter to be thrown out of the system, and nerves, which supply vitality to all parts and establish a bond of union with the whole. Insensible as the bones appear, they have all these essentials of living, organized matter, and they are constantly undergoing waste or decay, and renovation or growth, to which all living matter is subject. The proof of this is rendered evident by considering

that it is only by means of growth that they can adapt themselves to the varying necessities of the body. If they had not
life, the stature of the infant must have been that of the future
man, and when broken, by accident, they must have remained
forever disunited, and an encumbrance instead of an assistance
to the animal. In health, the bones are insensible to pain,
and receive ample protection in the soft and more sensitive
parts by which they are surrounded; their insensibility thus
enabling them to work without weariness or fatigue or pain.
But should accident break them, pain at once ensues, and, by
its guardian, protective power, promotes their recovery. By
its means inflammation takes place, vascular activity increases,
and thus a reunion of the broken parts is accomplished, and
that repose and quietude which are necessary for the parts to
adjust themselves to their healthy condition.

Let us now make a practical application of the principles we
have been considering to the preservation of health and the consequent prevention of disease and accident, to which we are liable. To be healthy, the bones, we have seen, must have a regular supply of good arterial blood, the arteries an abundance of
nervous force, by the nerves, and a proper balance between
the nutrient and absorbing vessels. These conditions must be
secured, or our bones suffer in their constitution and relations.
In the conduct of life, how are these to be attained? As the
bones are organs of motion, and endowed with hardness and
resistance, the frequent and regular performance of resistance
and motion is just as necessary to their well-being as air is to
the lungs, light to the eye, blood to the heart; accordingly
the bones become feeble, diseased and unfit for their uses, just
as the soft parts of the body do. This is a principle or law
of great practical importance, involving the law of exercise.
We will illustrate it further. No faculty of our mind, or organ

of our body, grows healthfully without *exercise.* This is nature's inexorable law.

Deprive any part of that exercise or action which nature
destined it to fulfill, it instantly grows *weak, diminishes in
size, shrivels,* and alters so much in appearance, that we
scarcely recognize it. Tie up the *large artery* that supplies
the arm with blood, and a change of structure begins at once,
and in the course of weeks what was once an elastic tube, becomes a ligamentious chord. Doom a *muscle* to inaction, and
it loses its bulk in a short time ; and, if continued for any
length of time, its power of contraction and muscular appearance is destroyed. All other parts I say obey this law. The
hard, apparently invisible bones, form no exception to it.
Experience has shown that *complete inaction* not only diminishes the size of the bones, but injures their structure so much
as to render them easily cut with a knife. This, though an
extreme case, shows a principle on which Nature conducts her
operations. Where there is great, though not total depriva-
tion of exercise, we see the most serious and pernicious de-
rangements ensuing. Witness the *ill health, curved spines* and
deformed figures which result from the restraints imposed by
our school-rooms and fashionable attire, the condemned attitudes of modern education, and the numerous stinted growths
of our system of training childhood—evils which would never
exist under a sound physiological education. The bones must
be systematically exercised, as they are the organs of motion,
and they, like the muscles which move them, suffer without
obedience to this law of their healthful existence.

The period of early youth is a period teeming with activity
and life ; development and growth are hourly going forward,
and the nutritive system consequently is in a state of increasing
activity. Abundance of good, wholesome food is *now* indis-

pensable to health. Witness the keen appetites and vigorous digestion of healthy children.

To fulfill the indications of nature and obey this law, children must be supplied with that abundance of nutritious aliments which she demands—and accordingly we see, among the laborious classes, the children born in such times afflicted with rickets, the *bones* being soft, tumid and weak, which condition, if continued into manhood and womanhood, leads to deformities and distortions.

RESUME OF WHAT WE HAVE LEARNED FROM OUR CONSIDERATION OF THE BONES.

1st.. That the bones are organs of resistance and motion, and must be *used*, by giving them suitable *exercise*. That this gives them health, strength, and increased ability to endure labor.

2d. That bones have, penetrating their structure, arteries, veins, nerves and absorbent vessels; that the arteries carry the blood that nourishes them; the veins return it to the heart; the nerves are the sources of life and motion, and the absorbents remove the waste that is constantly going on from every organic structure endowed with life. In order that the bones should grow, therefore, they must be supplied with an abundance of wholesome food, especially at that period of life (early childhood) when growth is most active. That the vigorous appetite of healthy children point to this as Nature's law or plan for development, and to violate it is to entail upon the child irreparable injury.

3d. That the practice followed by many ignorant parents of continually soliciting young children to stand up or walk *long before* the bones have acquired sufficient power of resistance and the muscles sufficient power of contraction to enable the child to cope with gravitation, tends to produce *curvature* of the bones. We also see how hurtful *leading strings* must be, and how injurious the *indiscriminate* use of *dumb-bells* in *early life* must necessarily prove.

4th. That we cannot *strengthen* the body by the use of *stays*, nor by lying down for hours at a time. That there is no royal road to strength and health, and no method by which its enjoyment can be secured, while we dispense with adequate and appropriate exercise in some shape or form.

THE MUSCLES.

Almost immediately beneath the skin lie the muscles. Though constantly in use every moment of our waking lives, and indispensable in the performance of every act of usefulness, they are less known and less regarded than their importance demands.

As the consideration of the muscular system involves the *principles or laws* which regulate *exercise, symmetry of form* and *grace of movement*, a correct exposition of these principles will not only be of deep interest to the general reader, but prove of special regard to the parent and teacher, and all others interested in the welfare of the young. *Muscle,* commonly known as *meat* or flesh, is composed of distinct fibres or threads arranged in bundles, separated from each other by a sheath of cellular membrane. In the cells of this membrane fat is deposited, which gives the rounded, plump outline to the limbs. The cellular membrane being loose, allows of perfect freedom of motion to the muscles.

All voluntary motion is accomplished by the contraction of muscles, acting upon bones, by having one end attached to one bone by their *origin*, and the other end to another bone by their insertion, the first being the fixed extremity towards which the opposite or movable end, called insertion, is directed by the shortening or contraction of the middle or intervening portion, called the belly. Hence, all the movements of the body are brought about by this process of muscular shortening, and the true function of muscles is contractility, or the power of shortening their fibres on the presentation of appropriate stimuli, and of again relaxing them when the stimulus is withdrawn. While the *chief* purpose of the muscles is to carry out our volitions by acting on the bones, they at the same time answer *other* highly important ends. They aid in the circulation or distribution of the blood, and prevent its undue accumulation in the central organs of the body. The important functions of *digestion*, or rather assimilation, respiration, secretion and absorption, are also carried forward, and the health of the entire body thus established. In short, without this invaluable characteristic of muscular fibre, life with all its attendant blessings could not for one moment exist.

ON WHAT HEALTHY CONTRACTILITY DEPENDS.

1st. The muscle must be strong and healthy originally. *Size* is a measure of power (all things being equal); small muscles, feeble power—large muscles, strength and vigor of action. This is the rule. *Use* a muscle much, and you increase its size and efficiency for work. Doom it to *inaction*, and its fibres shrink, grow feeble and inefficient. The reason is by *action* more *arterial and nutritive blood* is drawn to its fibres, and therefore more strength and force is manufactured. By *inaction* the opposite result is attained. Hence the inmates of large factories suffer so much, and those of our female boarding schools. By continued labor, long confinement and poor, scanty diet, the muscles become feeble, stinted in growth and injured in their structure, and the blood rendered poor by innutritious and insufficient food and by bad air, is incapable of repairing the waste daily going on from exercise. Languor, debility and exhaustion follow, and the mind is liable to be assailed by the stimulus of ardent spirits, or the seductions of reckless animal passions. The continued growth of early years demands highly nutritious food. Deny this and you inflict an injury on the health which no subsequent treatment can relieve.

2d. After strong, healthy muscles, there must be provided a proper stimulus through the nerves, or else the muscles are inert, inactive. From the mind comes will; the nerve conveys this will to the muscle, and the contraction follows its dictates. Thus, energy of mind brings about energy of muscle—feebleness of mind, weakness of muscle. Hence, as voluntary motion requires nervous power as much as muscular fibre, whatever disturbs the action of the nerves affects the quality and quantity of motion. Injuries of the brain are known to cause palsy or want of power in the muscle; narcotics and sleep suspend voluntary motion; ardent spirits also disturb regular muscular movements, and cause the unsteady gait of the drunkard.

WHAT EVILS FOLLOW MUSCULAR INACTIVITY.

Healthy muscular action consists of regular alternate contraction and relaxation of its fibres. Permanent contraction is therefore mischievous and impossible. Doom any one to stand in one position, and how weary and exhausted he soon becomes. Extend the arm with a ball in the hand, and how soon the painful attitude causes the hand to droop. Hence, great mischief is done in the education of our sons and daughters by forced confinement to one position or a limited variety of movements. This error of our fathers is now beginning to be recognized, and we are happy to say a more auspicious era is dawning in society. 'Till Nature's laws are recognized in the growth of our minds and bodies, evils will continually affect the race, nor will our kind mother cease to punish till we cease to sin.

WE SHOULD COMBINE MENTAL WITH MUSCULAR EXERCISE.

Movements to be healthy should be *varied* and *pleasing*, not *monotonous* and *dull*. He who saunters along without an object, grows weary. A walk, how listless and unprofitable, when taken against the inclination and merely for exercise. "In the one case the nervous impulse is full and harmonious; in the other the muscles obliged to work without that full nervous impulse which Nature has deemed to be essential to their healthy and energetic action."

The poet hath said wisely :—

> "*In what ere you sweat*
> *Indulge your taste.* Some love the manly toils;
> The tenis some, and some the graceful dance.
> Others more hardy, range the purple heath,
> Or naked stubble, where from field to field
> The sounding covies urge their laboring flight,
> Eager amid the rising cloud to pour
> The gun's unerring thunder; and there are
> Whom still the meed of the green archer charms.
> *He chooses best whose labor entertains*
> *His vacant fancy most.* THE TOIL YOU HATE
> FATIGUES YOU SOON, AND SCARCE IMPROVES YOUR LIMBS."

HOW EXERCISE AFFECTS THE ORGANS OF OUR BODY.

This is a deeply interesting topic, and concerns all classes of society. When any part of the body is put in action, the two processes of waste and renewal take place with greater rapidity and in due proportion to each other. To meet this condition, the arteries and nerves carry forward more blood and nerve power, and nutrition and nerve energy become greater. As soon as the action ceases the excitement subsides, and the blood and nerve force return to their original state.

Renew, now, the exercise at moderate intervals, frequently, and the increased action of the nerves and vessels proves more permanent. *Nutrition exceeds waste*, and the part gains in size, vigor and activity. Let the exercise be repeated too often or carried too far, so as to cause fatigue and exhaust the vital force ; *waste* then exceeds *nutrition;* the part loses in volume and power, and a painful sense of weariness, exhaustion and· fatigue comes on. Should exercise, on the other hand, be entirely neglected, the life of the part decays or grows feeble, from the absence of nerve power and the energizing effect of the arterial blood being withdrawn. Tie up a limb, and how soon its energy departs and every motion then becomes painful. So it is with the body when exercise is neglected. It grows weak, dull and unfit for energetic effort; all the functions languish. Regular systematic exercise, on the other hand, suitable in kind and degree, yields a grateful sense of comfort and activity, and we feel ready for the performance of every duty, whether of the mind or of the body. . If the exertion be excessive, painful weariness and exhaustion comes on, not relieved by rest, and which prevents sleep for a long time. Over-fatigue in riding, walking or hunting, in the same way induces feebleness and restlessness, which prevents sleep on lying down; and the person on rising in the morning feels languor of mind and feebleness of body, which continue till the exhaustion is removed, and then repose of a sound and invigorating character may be enjoyed,

RESUME.

RULES FOR ALL, TO PROMOTE THE HEALTHY DEVELOPMENT OF THE MUSCULAR SYSTEM.

First. Proportion the exercise to the strength and constitution, and be sure not to carry it beyond the point at which waste exceeds nutrition, and exhaustion takes the place of strength.

Secondly. It should be taken regularly, after proper intervals of rest, to insure the permanence of the healthy impulse.

Thirdly. Take an interest in whatever you engage to do. "For the labor we delight in physics pain."

HOW EXERCISE IMPARTS STRENGTH AND TONE TO THE REST OF THE BODY.

The foregoing remarks have shown us how *exercise* strengthens the muscular system. We will now endeavor to point out how the rest of the body acquires *health*, *strength* and *vigor* from judicious activity. Every part of our bodies, but especially our bones and muscles, declare in the most unmistakable manner that we are made for *action*. Every atom in the universe is ceaselessly active; a state of rest would destroy that harmony that everywhere prevails. Without industry the social state of man would relapse into barbarism. Now, this activity, energy or life, is solely maintained by the circulation acted on by the nervous force from the brain. Without *blood* we could not live. It is the " life of all flesh." Growth and strength depend on it. It is from the *blood*, which is but transmuted food, etc., that every fibre and atom of our bodies are made. First, then, you will observe, in following the food from the mouth through all its intermediate changes, until it has become blood, that almost all those changes are wrought upon it by the agency of the several fluids, juices or secretions which it meets with in the mouth, stomach and bowels ; and that consequently its due conversion into healthy blood depends upon the healthy quality and abundant quantity of these secretions. But these secretions, like everything else in the body, are formed out of the blood, and their quantity and quality will consequently depend upon the quantity of *red blood*, wherewith the organs in which they are produced are supplied. And the quantity of blood with which these organs are furnished, must depend upon the vigor and activity of the heart and arteries, whose office it is to convey it. Thus it is clearly manifest, that a vigorous circulation is absolutely necessary to the assimilation (vulgarly called digestion) of our food. Whatever causes and habits of life, therefore, are calculated to give strength and activity to the circulation—as, for instance, *exercise*,—are clearly of the first importance to the nutrition, and therefore to the health and strength of the body ; and whatever causes and habits have a tendency to depress the energy of the circulation, to allow the blood to creep languidly through the body, instead of dashing along its channels cheerily and energetically,—as, for instance, cushioned laziness, which rides when it should walk,—must of necessity have the direct effect of impairing assimilation, and therefore of enfeebling the strength and

sapping the foundations of health. Hence, also, it follows
directly that whatever causes are calculated to increase sensibility, to make us tender, have an immediate and powerful
effect in impeding the conversion of our food into blood, and
therefore of impeding the process of nutrition. Hence the
mischief of a daily indulgence in what are miscalled the comforts of life, but which are, in reality, most pernicious and
unnatural luxuries. A few of these are table indulgences,
lounging on couches, warm carpeted rooms, window curtains,
bed curtains, blazing fires, soft beds, flannel under-clothes, (I
speak of the healthy, not of the sickly invalid), novel reading, hot suppers, and the last, by no means the least, that
precious piece of foolery called passive exercise ; that is lolling,
at ease, in a stuffed and cushioned carriage. Not that I would
totally abolish any one of these, except, perhaps, hot suppers
and soft beds, but that I wish, by proving to you their evil
influences, to induce you to use them as sparingly as the con-
ventional habits of society will permit. Though I confess, for
my own part, I see no reason why any one should feel himself called upon to injure his health, to blur the beauty of
God's noblest work, solely to gratify the capricious whim of
that many-headed master, that blatant beast, called *Society*.

Finally, says a learned physician, "you will observe that
everything connected with life, all the *actions*, the *energies*
and *beauties* of the body, all the *actions*, the *energies* and *beauties* of the mind as well as the mind and body themselves are
under the dominion of the circulation of the blood, from
which both mind and body must inevitably derive each its
tone and character. So that the body and the mind are like
a jerkin and a jerkin's lining—rumple the one and you rumple
the other.

THE BEST TIME FOR EXERCISE.

1st. So long as we are healthy, exercise may be taken at almost any time, *except immediately after or before a full meal.* The invalid, and all those not robust, it behooves them to be more circumspect. The system must be able to meet the demands of activity, in order to derive benefit from it. This is generally the case about two hours after a moderate meal. The forenoon is therefore the best time. From 1 to 2 o'clock, P. M., the energies begin to decline, and all active exertion should take place *before* these hours.

2d. To secure the full benefit of exercise, it should always be taken in *pure air*, and when the weather permits in the *open air*, and should interest and occupy both mind and body.

3d. The garments worn during exercise should be loose, so as to admit of that freedom of muscle which is essential to its healthy activity and growth.

KINDS OF EXERCISE.

As the object of exercise is to develop the health, strength and beauty of the whole muscular system, it should be often varied and suited to individual peculiarities. *Walking* is perhaps the best form in which activity can be sought. Here mother Nature invites her children to her bounteous table. The fruits and flowers, the birds, rocks and plants, present a most attractive and profitable invitation to all in pursuit of health.

Riding is admirably adapted to dyspeptic and nervous invalids, and for certain stages of pulmonary disease.

Dancing, when not practiced in heated rooms and at late hours, and in fashionable attire, is eminently a graceful and healthy amusement. *Dumb-bells*, when adapted to the constitution of the person, and under professional physiological advisers, are full of strength and health. The same remark applies to the various forms of gymnastic and callisthenic exercises. *Fencing* is both useful, graceful, manly and invigorating. Its *abuse* is the only objection to its indulgence; knowledge of the body must regulate its exercise. *Reading* aloud, *recitations* and *declamations* are exceedingly useful to the body, when practiced under the restrictions of constitutional adaptation. They should in no case induce fatigue. The intelligent reader who has gone with us thus far upon this deeply interesting and important topic of *exercise*, must have come to the conclusion that there can be no perfect health and strength without exercise of both body and mind. The end of man's existence is alone attained by it, and the experience of all ages testifies to its efficacy.

THE SKIN.

In the mechanism of that great masterpiece of Divine wisdom (our body), no one of its elementary parts is deserving of more consideration than the skin. In a hygienic point of view, no structure operates more directly on the health of the body than the condition of the skin—and if popular ignorance is to be a measure for guiding us in an exposition of its functions and laws for its healthy action, none is deserving of more attention than the skin. Spread over the surface of the body, and *binding* together and *protecting* from harm the more delicate structures beneath it, is that membraneous covering called the skin. Anatomists speak of *three* layers entering into its composition, each performing *distinct* functions. The first layer from without inwards is the epidermis, *scarf skin*, or *cuticle;* the second, the mucous coat; the third, the true skin.

The cuticle is considered devoid of sensibility, for it has neither nerves nor blood vessels, and neither bleeds nor feels pain; whether porous or not is perhaps a matter of doubt, but both absorption and exhalation take place through its surface. Its structure is in harmony with its use; as an insensible surface it *shields* the nerves of the true skin beneath it, and thus *protects* them from the friction and rough contact of material things; by obstructing evaporation it keeps moist and supple the true skin, and by opposing a barrier to the absorption of numerous poisonous agents it is frequently exposed to, it protects the life and health of the body from that destruction it would otherwise be inevitably doomed.

Next in order comes the *mucous coat.* It likewise is devoid of sensibility and blood, and seems to act simply as a shield to the nerves and vessels of the true skin. Finally comes the *true skin*—dermis or corion. Here we have an organized structure of vast importance, full of life, and therefore having arteries, nerves, veins and absorbents. It is the seat of *perspiration*, on which so much of our life and health depend, and the residence of the nerves of sensation or touch. Four distinct functions are performed by it. First, it throws out of the system *waste matter;* secondly, it regulates the heat of the body; thirdly, it takes up or absorbs; fourthly, it feels, or is endowed with sensation or touch.

It has been stated before that every part of our bodies is constantly being wasted or destroyed, and then replenished or renewed, and that these two functions are maintained through the arterial blood. The skin being bountifully endowed with red blood, is thus admirably adapted for exhaling waste or hurtful matter that nature is constantly throwing out of the body. This waste matter is familiarly known as perspiration or sweat, and of its hurtful character when suddenly checked yea, I might say destructive (for death does sometimes result from its suppression), everyday life offers many examples.

Under ordinary circumstances, the matter of perspiration is so insensibly carried out that it is scarcely perceptible, but nevertheless, when interrupted by *sudden cold* is productive of the most serious mischief to the functions. All who value health, and specially the delicate woman, must look to this great destroyer of their constitution. An Italian physician by the name of Sanctorious, weighed himself, his food, &c., for thirty years, and came to the conclusion that five-eighths of what was taken into the body passed out through the skin, leaving three-eighths to escape by the lungs, kidneys and bowels. Taking the lowest estimate of the French chemist, Lavoisier, we find the skin removing from the body *twenty ounces of matter* daily—thus furnishing ample reasons for the mischief accruing when this important function is arrested. Rheumatism, neuralgia, coughs, colds, consumption, diarrhœa, dysentery, monthly suppressions in woman, croup in young children, and fever, are among the ills that flesh is heir to when perspiration is suddenly checked. It is only necessary for us to consider that for every twenty-four hours this function of perspiration, whether sensible or insensible, is retarded, we have twenty ounces, at the lowest estimate, of poisonous matter oppressing and deranging the functions of life. Is it to be wondered at that men, women and children have their health ruined from *cold?* On your peril, then, take care of your precious body !

RESUME.

LAWS OR RULES FOR OUR GUIDANCE IN PRESERVING THE
HEALTH OF THE SKIN.—These rules follow necessarily from
the constitution and relations of the skin as unfolded in the
preceding remarks. They have their foundation in *nature*,
and do not owe their authority to the opinion of any man or
set of men :—

1st. The waste matter thrown daily to the surface, must be
removed daily by proper washing.

2d. A *sufficient amount* of clothing of a suitable texture
must be worn, to prevent an abiding sensation of chilliness ;
not too much, but just enough. Experience here must guide.

3d. All effort to *harden* children by exposure to cold is unnatural, and results in disaster and frequently death.

4th. The clothing must be loose, not tight, and changed at
least once in two weeks.

5th. As to the nature of the material which should be worn
next the skin, it is difficult to express a decided opinion.
Some prefer flannel ; some the Canton flannel, and some silk,
and others cotton. To the hardy and much exposed, simply
cotton is, perhaps, alone desirable. To the more delicate and
fleshy, woolen texture may be needed. A wise, careful observation of our feelings must guide ; a warm skin must be
had, and this consideration is to be our guide.

6th. *Light* and *pure air*, are indispensable elements to the
health of every part of the body. We must therefore avoid
dark rooms and ill-ventilated apartments if we desire warm,
vigorous skins.

BATHS AND BATHING.

In connection with the skin we must append a few remarks on this engrossing topic.

Man being designed for a life of activity, he must perspire, to be healthy ; and this perspiration having within it the elements of hurtful, poisonous matter, must be removed. Now, as to the *form* of washing, and *frequency*, much difference of opinion prevails—whether we should sponge daily, or bathe daily ; and whether the water should be cool, cold, tepid or warm, are the questions for our decision. The observations of an eminent Scotch physician, Dr. A. Combe, are here worthy of all consideration. For *general* use he says the tepid or warm bath seems to be more suitable than the cold, especially in *winter*, and for those who are not robust and full of animal heat Where a warm glow over the surface does not follow the use of the cold bath, its use is forbidden, for it does harm. In *summer*, however, when no constitutional peculiarities forbid, the cold bath may be used with decided benefit. Before breakfast or immediately after fatigue, when the power to *react* is feeble, cold bathing is decidedly hurtful. But for those who are not robust, daily sponging of the body with cold water, or vinegar and water, or salt and water, is the best substitute for the cold bath, especially when proper efforts are made to get up a glow on the skin by friction with coarse towels, an active walk, or appropriate gymnastic exercises. The temperature of the bath should range from 85 degrees to 98 degrees, and the duration of the immersion may vary from fifteen minutes to an hour, according to circumstances.

A person of sound health may take a bath at any time, except immediately after meals. But the best time is the forenoon, two or three hours after the morning meal. Many imagine the tepid and warm bath weakening, but experience has shown the fallacy of these views. "If one-tenth of the persevering attention and labor bestowed to so much purpose on rubbing down and currying the skins of horses, were bestowed by the human race in keeping their skins in good condition, and in proper attention to diet and clothing, colds, nervous diseases, and stomach complaints would cease to form so large an item in the catalogue of human miseries."

THE LUNGS.

A consideration of the Lungs involves a consideration of the circulation, in order to understand their uses and adaption. The purpose for which we breathe or respire, is to introduce that element of the air known as *oxygen*, by means of which the blood is purified and rendered fit for its work of assimilation or growth of the solid tissues of the body, and for the further purpose, it is supposed, of introducing this vital element, oxygen, into the blood, to be carried by it to all parts of the body, to complete the work of waste or breaking down, which is incessantly going forward throughout its entire structure. We must then have clear ideas of this circulation before we proceed further. There are two kinds of blood in our body; the one known as *red, crimson*, or arterial blood; the other, *dark purple* or *venous blood*. The first, only, is the life-giving or nutritious blood; the second innutritious, or life, destroying blood. Starting from the left side of the heart it enters a great artery called the aorta, and is distributed by it and its subdivisions to every part of the body, When the work of nutrition or repair has been finished, the blood, altered in its color and properties, is returned in the contrary direction, to the right side of the heart. It is now venous or purple blood, and is totally unfit for sustaining life. Two things are now necessary before the blood can enter on its benevolent life-giving mission. It must be supplied with *new matter* from the food, called chyle, and with pure oxygen from the lungs. The first is supplied by the chyle, which is forced into the venous blood just before it enters the right side of the heart, and the second by the process of respiration, which takes place in the air-cells of the lungs. When the venous blood enters the right side of the heart, it is driven by its contractile force into a large artery, called the Pulmonary, which leads directly, by separate branches, to the two lungs; here it meets with the oxygen in the air, and is changed into arterial blood; it is then conducted to the left side of the heart, to be again sent on its merciful errand of imparting life and energy to the whole body. The importance of this process cannot be over-estimated. Upon the fidelity with which it is performed depends the life, the health, the energy, the vigor, the strength, and beauty of the whole human being. In Great Britian nearly fifty thousand perish annually from consumption. In New England consumption is the disease of its people. And in the Great West its ravages are becoming more extensive than those of any other malady. Faulty respiration is the cause. Remove the cause and its legitimate effect will soon cease.

CONDITIONS FOR HEALTHY RESPIRATION.

1st. *An original soundness of the lungs.* This comes through descent; the ancestor pre-determining the lungs of his children. No fact is better established than the descent of tuberculous or pulmonary disease from parent to offspring.

This is its chief means of perpetuation. Regulate marriages on sound physiological principles, and you arrest the ravages of consumption.

2. The blood must be healthy. Poor blood, bad growth, feeble development, and unhealthy lungs, all go together. The children of the poor and badly-fed are more subject to consumption than those of the rich and well-fed. Feed an animal on food not nutritious, and it is said tubercles will be developed on the lungs. Among the higher classes, when consumption prevails, the cause is to be looked for, (when not hereditary) in bad habits—generally excessive eating, neglect of the bath, bad air, and imperfect digestion of their food.

3d. Anything which impedes the free expansion of the lungs favors consumption. Hence stays, tight garments over the chest, heavy dresses suspended from the hips, stooping, confined attitudes, all injure the lungs and favor the develop- . ment of consumption.

4th. Such is the influence of the mind on the health of the body, that the indulgence of grief, sorrow, fear, anxiety (all of which are vicious in their influence on life), is powerful to predispose to consumption; while hope, joy, benevolence, love, sympathy, exilarate and improve the tone and vigor of all the functions.

5th. *Pure, fresh air.* This, where all other conditions are necessary to the healthful working of the "machinery of our bodies" is indisputable. We have already seen that *oxygen* is the pabulum of life; the great prime mover of the machinery of our bodies; that it acts directly on the blood by removing its impurities, fitting it for nutrition; and indirectly on all the tissues of the body, by oxidating, as the chemists say, the elements of waste. The *life force*, which resides pre-eminently in the brain and nerve centers, could not be developed without oxygen ; hence the prime value to the lungs of pure air. How this oxygen acts in effecting its changes is a subject of dispute, but this much all agree on, that when the venous blood reaches the lungs it is *unfit for life* ; that after coming in contact with pure air in the air-cells its color is changed, and it instantly takes on the red arterial hue, and this change is owing to the oxygen. Hence it is called *vital air*.

THE MEANS OF SECURING PURE AIR OR VENTIL-ATION.

Each individual breathes on an average from fourteen to twenty times in a minute, and inhales at each breathing forty cubic inches of air. Taking the lowest estimate at twenty inches, and the number of respirations at fifteen, in one minute each person consumes or needs 300 cubic inches of pure air, and in one hour one hogshead. This should be the basis of every plan for securing pure air for the lungs. During this hour, it is estimated that 1440 cubic inches of oxygen disappear by entering the blood, and an equal amount of poisonous carbonic acid gas takes their place, thus furnishing a frightful source of atmospheric impurity. Could we know all the ills inflicted on humanity from impure air, the catalogue would appal us by their number and magnitude. Our private residences, churches, halls, school-houses, steam-boats, halls of legislation, hotels, manufactories, and all places where human beings congregate, present not one aspect that indicates that human beings need pure air, so wretchedly are they contrived. At present our chief reliance is on the door and window, and these would pour such drafts of cold air, in winter, on the surface, that to open them would endanger our lives, or subject us to the perils of colds and rheumatism.

The plan adopted in the Edinburg Infirmary is the best, both for public and private edifices. Fresh air is introduced by large circular openings in the floor, and the vitiated air escapes by similar openings in the roof. The contrivance admits of the air being heated in winter before it enters the wards, and thus all danger from cold currents is avoided.

HOW OUR BODIES ARE HEATED.

We all know the importance of heat to life, but it is not generally known *how* the heat of our bodies is produced, and the conditions on which the healthful production depends. Did the body not possess such a power, our long winters would convert our blood into a solid mass and destroy life. A most instructive relation is known to exist between the size and vigor of the lungs and the quantity of animal heat. The larger and healthier the lungs the greater the quantity of heat, and *vice versa*. *Birds* furnish a striking illustration of this relation, their power of generating heat being remarkable, and they possess the lungs in the greatest perfection.

The nervous system, or brain, exerts a marked influence over the warmth of our bodies. Grief and anxiety depress the circulation, and cause slow breathing, and imperfect assimilation of the food, and coldness of the extremities; cheerfulness, joy, hope and love warm up the whole body, and quicken the circulation, &c. Good food and active digestion, with active exercise in the open air, develops greatly the warming power, while indolence obstructs it and causes chilliness. The activity of the circulation, and the introduction of oxygen into it through the lungs, are the main conditions for healthy warmth.

HOW TO EDUCATE THE LUNGS.

First and most important is the proper *exercise* of the lungs. This is the means Nature has appointed for developing and strengthening all of her organs. The lungs are designed to be used, and if habitually inactive, they grow feeble and remain small; or, if exercise be *excessive*, or at *improper times*, disease invades that structure. Two kinds of exercise are worthy of consideration—*direct* and *indirect*. Crying, singing, speaking, reading aloud are examples of the first. All kinds of muscular effort which cause a deeper and quicker breathing, such as rowing a boat, fencing, quoits, shuttle-cock, dumb-bells and gymnastics are of the second. If feeble lungs are inherited, our *earliest* attention should be drawn to their development and strength by *habitual* use of both the above-mentioned kinds, in conformity with the laws regulating the use of muscles. Climbing a mountain in pursuit of interesting scientific objects is highly beneficial.

This caution, however, must always be observed before any attempt at exercise be indulged—*no active disease must be present in the lungs.* Then both direct and indirect exercise may be practised to the extent of perspiring daily. From puberty to twenty-one is the critical period with those born with a consumptive tendency. Nature is now engaged in building up her temple, and her nutritive forces are active and demand attention. Exclusive attention to bodily health should now engross the subject, and all close and confining pursuits carefully avoided. Students should abjure books and indoor work, and take to the fields, the mountains and the ocean. When the critical period has passed, the mind will then resume its activity, and may be indulged with safety in regulated and systematic study.

In concluding this deeply important and interesting topic, let me say that the oxidation of blood, which is accomplished through the lungs, is the most important, where all are important, of all the living actions.

THE BRAIN, &c.

Man was given dominion over the fishes in the sea, the birds of the air, and the animals of the earth. This dominion he is enabled to use, preserve, control and extend through his endowment of a superior mind. In past ages of the world, and even with many minds of the present, the inquiry has been keenly pressed, "What is the nature of the mind?" but no satisfactory answer has yet been given. It has also been earnestly and zealously discussed, whether the mind acts independently of organized matter, and whether it acts as one lower or has a plurality of powers, and whether these separate and distinct powers act through and are dependent for their manifestations in this life on separate and distinct portions of the brain? I refer to these inquiries, not simply as deeply interesting and practical questions, as they are, but because they have a *direct bearing* on the subject of Hygiene.

When men become more enlightened, and less under the dominion of tradition or the authority simply of man, and more under that divine teacher, Nature, or Nature's God, it will be universally taught in our schools and colleges, and preached from our pulpits: That the mind of man acts through the brain, and is dependent in this life, for its power to act, upon that brain and its tributary nerves ; and that upon the healthfulness of that brain and those nerves depend the health and strength of the entire body. Therefore, keep the brain and nervous system healthy, and you keep the body healthy,—a doctrine of more practical value than any other connected with the domain of physiological science. Keep in mind, then, and let it be deeply engraven on the tablets ot your memorial power, that "activity of mind and activity ot brain are inseparable, and that every change in the one is attended by a corresponding change in the other."

If we use *stimulants*, and highly excite the *brain*, the *mind* will be disturbed in an equal degree. If by the sudden indulgence of *passion* the mind be roused, the vessels of the brain will be distended with blood, and redness will suffuse the face, and excitement of the brain will show itself in the most distinct manner, as though it were produced by a physical cause.

Let us, then, inquire into the *laws* on which this healthy association of mind and brain depends, that we may yield them a ready obedience, and thus escape those *innumerable ills* that now flood society, and threaten at some future day to merge it in a sea of ruin.

We have seen elsewhere that the body, to enjoy health, must, 1st, have an *original* soundness which it gets from its parents or ancestors ; that all parts of the body are bound together by the nervous system, and that the same blood which develops one part develops every part. The brain, then, to be healthy, must have this ancestral vigor, or all attempts to strengthen will be feeble and imperfect ; the *mother* here plays the most important part, transmitting more of good or ill to the child than the *father*. 2d. The brain must be supplied with an abundance of healthy arterial blood. Persons who live in close, ill-ventilated rooms, and our schools and public assemblies, where ventilation is disregarded, feel listless, apathetic, dull, frequently dizzy, and sometimes afflicted with headache and fainting fits, all of which grow out of the condition of the blood on the brain, as regards its oxygen. 3d. As man was made for activity and usefulness, he must use his brain powers for the benefit of his fellow-beings. *Exercise* of his faculties is then essential to their welfare, and through them to the welfare of the whole body; inactivity, the bain of earthly bliss. Persons who retire from the field of active labor prematurely, under the expectation of being happy, soon find their expectations sadly disappointed. The cause is to be sought in the *non-exercise* of their brains in reference to their fellow-beings. While we fully recognize the value of action as imperious, for true health and happiness, we must understand that regulated activity is always meant. A brain kept constantly on the stretch, in whatever pursuit, is sure to be invaded by inflammatory disease—apoplexy, insanity and paralysis. Dickens, Sir Walter Scott, Whitehead, Romilly, Castlereagh, Canning, and many other bright men of talent, urged on by ambition, were suddenly cut short in their career by the *inordinate action* of their brain.

> " The inordinate cup is unblessed,
> And the ingredient is a devil."

A great lesson is here taught to the young, ambitious students. They exhaust their brain force by continued exercise, and end finally in ill-health and the disappointment of their hopes. Said I not that the doctrine I am teaching, of mind being dependent on brain, and subject to the laws of organization, was deeply practical to this and succeeding generations? If the growing sons and daughters hope for enjoyment, usefulness and long life, they will make this doctrine the corner-stone of their earthly system of faith and practice, and give it a prominence in the practical affairs of life commensurate with its importance. .

HOW SHALL I EXERCISE MY BRAIN?

As we cannot think well and digest our food well at the same time, we must not attempt mental labor immediately after a full meal. This would impair both processes, and give us feeble thoughts and poor digestion. Hence literary men, who live by the labor of their brain, and men engaged in the engrossing cares of business, must rest at least one hour after meals before entering on the duties of their calling. Pleasant, agreeable conversation, or any amusing pastime at such a time is the object to be secured. The dyspepsia of students and literary men, &c., is traceable chiefly to this source.

WHEN SHALL I INDULGE IN BRAIN WORK?

Sir Walter Scott, it is said by his biographer, performed the greater portion of his literary work from 4 in the morning to 10 o'clock, embracing a period of six hours of that portion of the twenty-four when the mind and body are best prepared for intense work. This was wise, and may serve as a rule for the guidance of similar workers. After noon of the day the strength declines, indicating clearly the necessity of a decline of vigorous labor. I think it may be safely affirmed that men who desire a long, healthy life, should not work at night, but use the period of darkness for rest and such amusement as may prepare the brain for that repose which is essentially necessary for the healthful restoration of the bodily functions. There is a time set apart by Nature for all things. "Leaves have their time to fall and stars to set." The seasons succeed each other regularly, and seed-time and harvest have their periods. Periodicity of action seems to be a *law* The nervous system has its laws also, indicating a time for the successful performance of its duties. It tends to renew its labors or mode of action at stated periods, and hence *regularity* or *periodicity* is important in exercising the intellect or moral powers, thus laying the foundation for what we call *habits*. This periodical or *associated activity* is of vast importance in any attempt to improve our brains and bodies. It creates *dexterity*, *skill* and *facility* of action, and *increased power* to do what we may have to do. We do not thus improve the abstract, thinking intelligence by this process, but simply educate matter as a better instrument for the mind to work with.

OUR FOOD.

There is no question pertaining to the welfare of man, whether we regard it in reference to his health or his physical vigor, or to his intellectual and moral supremacy, that is more deserving of his earnest consideration than that which relates to his food. In determining *what kind* of food he shall eat, *how* it shall be cooked, and the *times* of eating it, his *reason* seems to have been little exercised—but rather the uneducated instincts of his nature and the necessities to which he may have been subjected by the locality in which his lot was temporarily cast. Hence, at one time his diet is exclusively vegetable, at another animal, at another a mixture of both vegetable and animal. Sometimes he prepares it with fire, at others it is eaten raw, and generelly eaten at such times as his convenience or necessities demand. That he has lived to a comparatively good old age upon every form of diet, and with comparative health, may readily be conceded with truth ; but that he has attained such a longevity, and such vigorous health, and such intellectual and moral vigor as he is capable of attaining under a diet regulated upon principles in harmony with his physiological nature, may well be doubted. With such *conflicting evidence* as that furnished by the *so-called* experience of the race, we might be perplexed how to decide. This much we do know, and affirm with certainty, that as the true nature of man is unfolded, so does his food change more and more to that of a fruit and vegetable character. That *animal* food is not increasing your strength is shown by the habits of prize-fighters, and porters, and men who tax their physical strength to the utmost. Even those who advocate the use of meat, restrict it to more sparing quantities than of old. Perhaps to the literary and sedentary it might safely be interdicted with great benefit to their physical and mental health.

The stomach is the organ for the reception of *food*, after it has undergone proper chewing by the teeth. Here it is acted on by a fluid called the gastric juice, and is converted into a mass called chyme. Hence it is forwarded to the first portion of the intestines, called the duodenum, and is here acted on by juices from the liver and pancreas, named bile and pancreatic juice. It is now taken to a large vein, which conducts it to the heart, from which it is propelled to the lungs for the purpose of being acted on by *oxygen* from the air, to complete its transformation into blood.

WHAT FOOD SHALL WE SELECT?

Shall it be animal, or vegetable, or a mixture of both? Guided by the practice of society, a mixed diet, we should say, of animal and vegetable food, at least once a day, is necessary for health and vigor. Taking the true nature of man, as deduced from his anatomical and physiological peculiarities, we should unquestionably adopt the fruit and vegetable system of food, as best suited to health, vigor, beauty and longevity. *Vegetable* food is more *digestible* than *animal; animal* is more *stimulating* than *vegetable*, owing to its containing more *nitrogen.* "The working people in almost every nation are obliged to live almost entirely on vegetable diet, because it is so much cheaper, for it takes fifteen times as much land to provide animal food as it does to supply vegetable diet. The working people of Ireland live on potatoes ; the peasantry of Lancashire and Cheshire, who are the handsomest race in England, lively chiefly on potatoes and buttermilk ; the bright and hardy Arabs live almost entirely on vegetable food ; the brave and vigorous Spartans never ate meat ; most of the hardiest soldiers in Northern Europe seldom taste of meat. From the creation to this day more than two-thirds of mankind never have eaten animal food ; and except in America, it is rare that the strongest laborers eat any meat."

One *principle* of great value to health should guide. Modern society has decreed that our tables should be furnished with *white* flour bread, instead of, the *brown unbolted* meal, thus robbing the flour of its bulky or innutritious elements—which, according to modern chemistry, supply the brain and nervous system and bones with the lime and phosphorus essential to their healthy action. This breeds a vast amount of evil, and is one cause of the bad teeth, small bones, and weak, feeble brains of this generation. Give us more coarse food for our daily consumption, and increased vigor will reward us for our obedience. To the adult and healthy we may say —If meat is chosen, let it be in *winter* and during the out-door active work of the day, and when spring and summer come, drop the meat (or use it sparingly) and live on the fruits and vegetables of the season. In regard to children it is safe to say that healthy milk is the best while the

teeth are undeveloped—after they appear the introduction of
solid food may gradually be made.

TOO MUCH OR TOO LITTLE.

No error is more fatal than that of *excess*. Better too little
than too much. When we put into our stomachs *more* than
nature calls for, by her sense of hunger, she uses what she
needs and leaves the *surplus*, to be thrown out of the body
through the lungs, kidneys and skin (thus over-working these
organs and causing exhaustion and irritation), or a deposit of
fat is made, which only serves to encumber. In all healthy
bodies there is a nervous condition which arises, and tells us
when we have enough. This plainly says, by its sense of indifference, *stop!* Variety and complexity stimulate the appetite to excess; simple food is the remedy. Condiments are
simply provocatives to the "too much." Alcott has properly
styled them "the medicine chest" of the table. We take
food to supply the waste daily going on in the body—and hunger is the sentinel to warn us when the food is to be taken.
Eat, then, for appetite, and not for the gratification of a few
nervous pappillæ spread upon the tongue. Regularity is important; the danger here is that we shall eat when the meal
time comes, and not the hunger. Americans hurry or bolt
their food, thus laying the foundation for imperfect assimilation. Eat slowly, and wait at least one hour after meals before
entering on business or actual labor.

OUR DRINK.

We drink for the purpose of supplying the fluid waste of our
bodies, or to relieve internal irritation. In a body perfectly

healthy we question whether the sense of thirst would exist; it seems to be a forced state. A properly selected aliment contains as much fluid as the health of the body demands. The habit of the American people in drinking with their meals and at intervals is detrimental to health. Tea and coffee are needless, if not detrimental, and every form of alcoholic stimulant is poisonous to the brain and nerves, undermines the constitution, and sows broadcast the seeds of disease. Water is the best fluid, and *oxygen* the only admissible stimulant.

THE BLOOD

"*The blood is the life.*" In itself it contains bone, cartilage, tendon, fat, muscles, membrane, nerve, brain; the most dense, the most volatile, the highest and lowest form of life. In the blood may be found all aliments, and the matter of all secretions. Whatever is digested or absorbed—food, water, poisons—is in the blood. From the blood is formed saliva, gastric juice, bile, the pancreatic juice, mucus, tears, sweat; and from it also are separated urine, fecal matter, and other secretions. The *blood*, moreover, is the direct source of the matter from which the female ovum is formed, as well as the fecundating spermatozoa of the male. Not less is the blood the source and continual pabulum of the highest organs of the mental and moral faculties. Shall we not keep the blood pure? Shall we not give to it a pure and simple aliment? Shall we make it feverish with flesh and stimulants and condiments, or poison it with narcotics? For be sure that pure blood comes from pure aliment and right habits of living; pure nutrition comes from a pure blood, pure secretion, and pure thoughts, feelings and actions. "For the pure all things are pure;" that is all things *must* be pure. The food must be pure ; the drink must be pure ; the air we breathe must be pure ; the whole body must be pure ; for purity is health and health is purity.

THE NERVOUS POWER.

Yet there is something higher than the blood, though of it, belonging to it, and nourished by it. This is the animal spirit, or nervous power. The nervous power, first of all, forms the blood. It presides over all of its secretions. It is by its intelligent and creative force, that from the same blood is made, here bone, there flesh, and elsewhere nervous fibre. It is by the directing influence of the nervous power that we have saliva from the parotid gland, *milk* from the mammary, *tears* from the lachrymal, *bile* from the liver, and urine from the kidneys. The *nervous power* is strengthened by exercise, and weakened by both indolence and exhaustion. Excess and impurity diminish its energy, and whatever deranges or impairs the nervous power, injures all the functions of life. Life is a collection of uses. What we do not use, we soon lose ; what we use too much, we also lose.

THE HARMONY OF LIFE.

The law of life is *harmony*. The perfect man is he who gives due exercise to every organ of his body, and every faculty of his brain. Thus harmony pervades all nature, and the true life of every organized being consists in the due and regular performance of all its functions. Thus harmony is *health*, and there can be no true health without this full and harmonious action.

Discord is disease. In a true life there must be the regular performance of all bodily functions, the exercise of the whole muscular system, intellectual pursuits, and enjoyments, and passional gratifications. These in their strength and variety make up the fullness of life.

HYGIENE—FOR CHILDREN.

Before we can rear a plant successfully, or an animal, whether it be an inferior or superior one, we must understand its *nature and its environment or surroundings.* This principle is recognized and acted on by the pomologist, florist and horticulturist, and the most perfect and beautiful fruits, flowers and garden vegetables spring up in obedience to its guidance. The Arab steed, the fleet greyhound, and the dull Conastoga pony; the delicious grape, the exquisite peach and the rich pippin, are the fruit of its teachings. No one, I think, at all conversant with organic nature, disputes its correctness. But strange as it appears, a principle so universally recognized when applied to animals and vegetables, is almost entirely ignored when the subject of our care is a little child. A law thus important must necessarily work the most intense mischief in its repeated violations, and we should accordingly expect to find in the number of children who are born into the world from year to year, the most extreme sufferers and a mortality the most appalling. Accordingly, between one-third and one-half of all the children born die within the first five years of their age. In England and Wales more than one-third of the total deaths occur under two years of age. In Belgium the mortality is equally great; one in every ten infants born alive is cut off within the first month; among the male children born in towns, a little more than one-half are alive at the end of five years. In Prussia, during a period of eight years, the mortality during the first year was in the pro-

portion of 26,000 to 100,000. In France it was as 21,457 to
100,000—not quite one-fifth; in Holland nearly one-fifth; in
Sweden the same.

Thus it appears, taking the average of the civilized countries
of Europe, in the midst of science and the comforts of civilization, where one would expect the treatment of the young to
be most rational, *two out of every nine* infants coming into life
die within the first year. Assuming the national standard of
a healthy lease of life to be three score and ten years, how
abundant and how active must be the seeds of death sown in
the infantile constitution or in its surroundings.

In New York City more than one-half the deaths are of children under five years of age. In Philadelphia, 51 per cent. of
the deaths of children were under five years. In the city of
Liverpool, in 1858, more than 26 per cent. of the whole number of deaths took place under one year of age, 38 per cent.
under two years, and 48 per cent. under five (being nearly
one-half). Two views present themselves to account for this
unnatural state of things :—

1st. Does this mortality form a "necessary part of the
Divine arrangements, which we can do nothing to modify or
prevent?" 2d. "Or does it proceed from the operation of
secondary causes, left purposely under our own control to a
very great extent, which may be partially obviated and rendered harmless by studying the nature of the 'infant constitution,' and adapting our conduct to the the *laws or conditions*
that regulate its functions."

That the first is not the true view is obvious from the fact
that just in proportion as ignorance of natural laws abound,
death abounds in childhood; and on the contrary, as we increase in knowledge of God's plans or purposes, as expressed
through our bodily organs, or as we obey these laws, we enjoy

a most gratifying immunity from suffering and premature death. The mortality, therefore, now, among the poor, in infancy, is much greater than among the rich ; in *towns* more than in the *country*. The second view remains the only just and natural one for the proper regulation of our conduct.

Let us inquire into the *laws* or *rules* or *principles* (for they mean the same thing) for the successful rearing of children.

LAWS FOR THE MOTHER.

1st. The *mother* must be healthy, and be surrounded with a reasonable share of the comforts of life.

2d. She must have a knowledge of the child's bodily functions, and of its relation to food, air, clothing, temperature and water ; for from the violation of these two laws proceed nearly all the sickness and early mortality in childhood.

3d. Habitual cheerfulness of mind, before and during pregnancy, and the avoidance of all unnatural sights calculated to alarm or terrify. It is acknowledged by the ablest observers and thinkers in the medical profession, that the mental condition of the mother affects directly the child. James I of England was alarmed at the sight of a drawn sword, which his mother Mary had beheld during her pregnancy. Hobbes, the philosopher, was excessively sensitive and timid, which he attributed to the fright in which his mother lived before he was born. Sir Walter Scott, in his "Tales of a Grandfather," tells us of the child of Lady Crowarty being born with the mark of an axe on its neck, because the mother labored for a long time under deep apprehension of seeing her husband brought to the block. A fit of *passion* in the nurse so affects the milk as to cause colic and indigestion in the suckling babe. But the most remarkable case on record of mental anxiety or sudden fear on the pregnant mother, is recorded by Baron Percy, an eminent French military surgeon, as having occurred after the siege of London in 1793. It is as follows : A violent cannonading had been kept up, which kept the women in a constant state of alarm. The arsenal at the same time also blew up with a frightful explosion. Out of ninety-two children born in that district a few months after, sixteen died at birth, thirty-three languished for eight or ten months and then died, eight became idiotic and died before five years, and two came into the world with numerous fractures of the bones of the limbs. Thus, two out of three nearly were actually killed through the mothers' alarm. Again, nearly all great men are descended from mothers remarkable for great intellect. ·

4th. The *habitual state* of mind and body, whether it be one of activity or inactivity, good or bad temper, ill or sound health, is indelibly stamped on the mind and body of the child. Through this law comes the character of our posterity and the health of the unborn generations.

5th. The diet should be plain and nourishing, accompanied by exercise daily in the open air.

6th. Cleanliness and tranquility demand the use of the *tepid* bath every few days. It keeps the pores of the skin open and the nervous system free from irritation, both positive elements of first-rate health. See that you cultivate them in this hour of your necessity.

LAWS FOR THE CHILD.

AIR—THE FUNCTIONS—FIRST RESPIRATION.—*First.* The child must have pure air, sufficiently warm and dry, and must have its clothing comfortable and loose, so as not to compress the chest. *Second.* The little child in almost an " instant is transferred from unconscious repose, solitude and darkness, to life and light and action. From being surrounded by a bland fluid of unvarying warmth, it passes at once to the rude contact of an ever-changing and colder air, and to a harder pressure, even from the softest clothing, than it ever before sustained. Previously nourished by the mother's blood, it must now seek and digest its own food, and throw out its own waste. The blood, once purified and restored through means of the mother's system, must be oxygenated by the child's own lungs. The animal heat once supplied to it from another source, must now be elaborated by the action of its own organs. Formerly defended from injury by the mother's sensations and watchfulness, its own nerves must now receive and communicate the impressions made by external objects. Through its smiles or its cries it must now announce to her ear and reveal to her judgment its safety or its danger, and if any of these important changes fail to take place in due time or order, its life may fall a sacrifice.

FOOD.

Previous) to the appearance of the teeth, the mother's milk is the best, in the absence of it a *healthty* wet nurse; next. fresh cow's milk diluted with half water and slightly sweetened, to resemble as nearly as possible the mother's milk. After the teeth appear, the use of broths, or beef tea, in connection with the milky and farinaceous diet, will be found to agree, always bearing in mind, that *excess in quantity or too frequent feeding are the evils to be avoided.*

CLEANLINESS.

The skin is an outlet for *waste or hurtful matter*, it must be remembered, exceeding the amount from both bowels and

kidneys, and it must therefore be removed daily with water
of the proper temperature, say 96 degrees, about the heat of
the child's body. This should be continued every morning;
never continued too long or used in a cold room. The soiled
garments must be carefully looked after and their place supplied with clean ones.

EXERCISE.

As the child is incapable of voluntary exercise, it must
therefore be accustomed to passive, for some weeks after birth;
this is done by carrying the child about the nursery in a horizontal position in the nurse's arms, and by rubbing the skin,
covering the entire surface of the body, with the hand. In
summer, the child should be as much in the open air as the
weather will permit, avoiding the morning and evening
chilliness. The above rules are for guidance during the first
weeks of infancy. As the child grows, its system acquires
strength and it shows active signs of muscular desire; then the
treatment must be changed. It wishes to use its arms
and legs, and no harm can result from the gratification of this
desire, properly guarded. The plan is to place the child on a
carpet and allow it free motion; all efforts at making the child
attempt to walk, by standing it on its feet, are decidedly
wrong, and only lead to mischief. The *North American
Indians* pursue this course with their children, and they grow
robust and are able to walk without help at six months old.
The great principle, says Dr. Combe, " in infancy, as in later
life, ought to be, to promote *self-regulated action, whether of
body or of mind, and to guide inexperience to the mode in
which nature intended the action to be performed.*"

SLEEP.

Growth and warmth are indispensable to the child, during the first four weeks, and in winter or early spring it must now sleep by its mother, and be, permitted to indulge its instincts. At the age of six weeks the child may be allowed a separate bed by the side of its mother.

THE BED.

Usage has decided the cradle to be the best, with its mattress and pillows, and such clothing as is comfortable and not oppressive.

QUANTITY OF SLEEP.

If the child be healthy, nature will determine this. Follow her indications. Periodicity is one of her *laws*, and we cannot too early direct the child to its observance in all of its movements, whether in eating, sleeping, or physical exercise, etc., etc.

TEETHING.

During the first six months of the child's existence, its entire dependence on the mother for its food renders it unnecessary to be provided with teeth; it accordingly has none, and draws its nourishment from its mother's breast. At six months, the first set of teeth appear; this is a natural process and not necessarily dangerous, and accordingly all *healthy* children pass through it without suffering. But · *delicate children* have to be carefully attended to or mischief will ensue. The organization of a child is suited by nature to its wants, and the kind of teeth bears a direct relation to the kind of food needed. Therefore we have different sets of teeth at different ages. When the food needed is watery and thin, and easily masticated, the jaws contain only twenty teeth, called *milk, or temporary teeth*, all developed before two or two and a half years. At seven, these fall out gradually and are replaced by a larger, stronger and more numerous set, called the *permanent teeth*. These are sometimes not completed till twenty or twenty-five years of age. The danger to the child during dentition arises from *irritation*, which may pass into inflammation and thence to death. Hence, the importance of great simplicity in diet, avoiding all excess or over-feeding, systematic *bathing in water of a suitable temperature, and the constant presence of pure air*. This last cannot be too much insisted on. If the weather is suitable, the child should be carried out daily into the open air, and when not fit, the room must have an uninterrupted circulation of unadulterated air adapted in temperature to the infant's constitution. Twenty-three thousand children perish annually in New York City, says Dr. Griscom, from want of pure air. It is the chief cause of infant mortality in our country, and the attention of mothers cannot be too earnestly directed to its pernicious influence.

MEDICINE FOR CHILDREN.

I cannot perform a more acceptable service to mothers than by introducing the following just and most sensible remarks from that excellent physician and philosopher, Dr. A. Comb :

" Many mothers are constantly administering medicines of one kind or another, and thereby deranging instead of promoting the healthy operation of the infant system. Instead of looking upon the animal economy as a mechanism constituted to work well under certain conditions, and having, in virtue of that constitution, a natural tendency to rectify any temporary aberrations under which it may suffer, provided the requisite conditions of action be fulfilled. They *seem to regard it as a machine, acting upon no fixed principles, and requiring now and then to be driven by some foreign impulse in the shape of medicine.* Under this impression, they are ever on the watch to see what *they can do* to keep it moving; and altogether distrustful of the sufficiency of the Creator's arrangements, they no sooner observe a symptom than they are ready with a remedy. Such persons never stop for a moment to inquire what the *cause* is, whether it has been or can be removed, or whether its removal will not of itself be sufficient to restore health. They jump at once to the fact that disease is there, and to a remedy for that fact. If the child is convulsed, they do not inquire whether the convulsions proceed from teething, indigestion, or worms, but forthwith administer a remedy to check the convulsions; and very probably the one used is inapplicable to the individual case, and both the disease and the cause being, in consequence, left in full operation, the danger, instead of being removed, is increased.

"This is no imaginary picture, but one of daily occurrence. Viewing disease as an entity lodged in the system, the uninformed and anxious parent hastens to expel it, and in so doing often perils the life of her child. When the truth comes to be generally known, that disease is not an abstract entity, but an aberration from the natural state of an organ or function, proceeding from some active cause, and is not to be removed till the diseased organ is again placed under the conditions essential to its healthy action, more attention will be paid to seeking the co-operation of Nature in our curative treatment, and much less mischief be done by rash attempts to expel the disease by force. The physician, when in his right place, is the 'servant and interpreter' of Nature, and not her ruler or opponent, and the same principle ought to apply with double force to the mother. Accordingly I have no hesitation in expressing my conviction, that a child can encounter few greater dangers than that of being subjected to the vigorous discipline of a medicine-giving mother or nurse; and wherever a mother of a family is observed to be ready with the use of calomel, anodynes, cordials, and other active drugs, the chances are, that one-half of her children will be found to have passed to a better world."

THE CAUSES OF BAD HEALTH.

Before we can remove an effect, the cause from which that effect flows must be recognized and removed. That ill health is a result, no one denies; but as to the cause of that result, a multiplicity of divergent and opposite opinions prevail. To entertain correct notions on this subject is a matter of first rate importance as upon their correctness will depend the kind of measures used for the purpose of overcoming the physical ills. Three views prevail :

1st. *Bad health* is looked upon as the result of causes beyond our control, and inflicted by the Ruler of the universe, as a punishment *for sin*.

2d. As the result of causes from which it is impossible to free ourselves—the accidents and contingencies of life.

3d. The direct result of the violation of those laws or conditions which govern the healthy well-being of every organ in our bodies, which laws it is within our power to know and obey.

If the *first* is believed to be true, all that we have

to do is to cease to sin in order to cease to suffer; we must become morally and religiously correct. If the *second*, nothing that we can do will avail to avert the calamities of ill health; we must patiently endure them. To the *third* cause we turn with some satisfaction, for here we have a rational foundation on which to erect a practical basis for human safety—the permanency and invariability of Law, discoverable through the only two sources from which our earthly knowledge comes—observation and experiment. Let us consult facts to sustain this *third* cause. In London, out of every one thousand infants born, six hundred and fifty die before reaching ten. In other places, half die before mature age. In New York, twenty-three thousand perish annually from bad air. No such mortality prevails among the lower animals, and an increased knowledge of the human body has enabled us to trace this unnatural mortality to bad air, bad food, exposure, etc.; causes within our reach and capable of removal. Again, the first 2350 years of the history of the race—that is, more than one-third of its acknowledged duration—"not a single instance is recorded of a child born blind, or deaf, or dumb, or idiotic, or malformed in any way." During this whole period, not a single case of a natural death in infancy, or childhood, or early manhood, or even of middle manhood is to be found. Not one man or woman died of disease. The simple record is, "and he died," or he died " in a good old age, and full of years;" or, he was old and full of days. No epidemic, nor even endemic disease prevailed, showing that they died the natural death of healthy men, and not the unnatural death of diseased ones. Bodily pain from disease is nowhere mentioned; no cholera infantum, scarlatina, measles, small-pox; not even a tooth-ache. So extraordinary was it for a son to die before his father, that an instance of it is deemed worthy of notice, and this first case of the reversal of natures laws, took place two thousand years after the creation of Adam. Rachel died at the birth of Benjamin, and this is the only case of puerperal death mentioned in the first 2400 years of the sacred history; and this took place during the fatigues of a patriarchal journey. One hundred years ago, when the pauper infants in London were brought up in bad air, fed on poor food, and crowded together in their workhouses, only one in twenty-four lived to be a year old. *Now*, under an improved system of treatment, the mortality is greatly lessened. In the country, how much less the rate of mortality than in the crowded city! In the days of the ancient Romans, their cities were frequently depopulated by plagues and pestilences. The present generation, by a more faithful observance of the organic laws, is now entirely exempt. In London, a like disregard of pure air, cleanliness and comfort, which was so fatal to the Romans, produced the same results; a destructive plague, which swept off its thousands and tens of thousands, till a great fire, by destroying the sources of impurity in those filthy homes, lanes and alleys, freed them from its deadly influence, and taught them that such calamities are not the wanton inflictions of a revengeful God, but the result of deadly emanations from their own filthy bodies and from the deadly and putrescent matter of their cess-pools.

Small Pox, before the introduction of vaccination by Jenner, annually swept off its thousands. Knowledge has now freed the disease of its terrors, and placed it within the perfect control of man. During the last century, *Ague* was so prevalent in Brittain that its attacks were looked upon as a necessary evil. By draining the land, removing filthy accumulations of matter, and the erection of comfortable houses, etc., the disease is now

unknown. A century ago, how different the health of the
sailor to what it is now! The famous expedition of Anson,
around the world from England, in 1740, numbering 1200 men,
lost, in twelve months, 865 men from fever and scurvy,
engendered by filth and foul air between the decks of their
vessels. Contrast this with the expeditions of Parry, of Ross,
and of Franklin. These distinguished navigators spent upwards of four years in the dreary regions of the North, with
scarcely any loss of life. Capt. Cook set sail from England on
the 13th of July, 1772, for a voyage around the world.
Towards the end of August, on arriving at the Cape of Good
Hope, only one man was on the sick list. Under similar circumstances, a Spanish ship reached the coast of Brazil with eighty
on the sick list, of whom twenty-eight died. The Resolution,
of Capt. Cook's expedition, performed a voyage of three years
and eighteen days, through all climates, from 52 degrees N.
to 71 degrees S., with the loss of only one man by disease,
out of 112. Such encouraging and gratifying results are
attributed to the constant care of his men, the great cleanliness
of his ships, and the admirable adaptation of their food to the
climates through which they passed, and the great cheerfulness
he encouraged and promoted among them. Three score and
ten are the years allotted for man, but few reach them, and no
prophet is needed to tell us that, until we live according to God's
laws, we shall continue to suffer and fall short of his intentions.

The average duration of life in 1844, was :

In Massachusetts,	33.74
England and Wales,	33.74
Prussia,	27.77
Sweden,	27.00
Russia,	19.18
Concord, (Mass.)	38.87
Dorchester, "	32.54
Plympton, "	40.80
Louisville, (Ky.)	17.87

In the most favored of these towns and countries there was an average s of 3-7 of life, and in the most unfavored more than 5-7.

In Massachusetts, among the farmers, the average age was 45 years, while among the poor it was only 27 years. In Concord, Mass., among the comfortable, the average age was 44 years, while among the poor it was only 24 years.

" It cannot be questioned, says Dr. Jarvis, that this depreciation of life is mainly chargeable to the general ignorance of the conditions of our existence on earth, and to a consequent failure to fulfill them. The knowledge of the laws of physical life has not been, and is not now considered, requisite for the conduct of our lives. Nor are the young instructed in these in order to prepare them to meet and avert the ' ills that flesh is heir to.' Physiology has not been included among the necessary studies of our schools, nor have men in older life thought it worth their while to attend to it. The remedy, then, for these evils and errors, is to incorporate the study and practice of Physiology in the course of universal education. Give this science a prominence in all our schools and colleges, in proportion to its importance, to its bearing upon human health and human life. Then will men and women be saved great suffering, and be so far prepared to fulfill their natural destiny on earth."

LAWS—The First Principles of Spiritual Liberty.

"Abundant evidences exist to the intelligent, conscious mind of the existence of a Deity, and of his attributes of power, wisdom and beneficence." But all this would be nothing but curious speculation if we did not hold the closest personal relations to that Being, and if our nature and destiny did not

primarily depend upon his attributes. If accident should cast
us upon the shores of an unknown country, our first inquiry
would be what kind of men and government lived and reigned
within its borders,—whether man-eaters or man-lovers ; whether
an arbitrary, cruel and tyrannical king, or a beneficent and pa-
rental one. So, when we are cast upon the shores of Time and
see this outspread universe around us, the most momentous
question we can ask is—what kind of a being governs it, and
what we must do to adjust ourselves to our new condition.
If God be the creator and governor of all existences, then he
is our creator and governor. If he be all-powerful, then it is
useless, nay, it is madness, for us to defy or resist him. If he
be omnipotent, then we cannot escape out of His jurisdiction,
as guilty men sometimes flee from the rod of the civil magis-
trate. If He be omniscient, then no omissions or technicali-
ties of an imperfect criminal code, no craft of counsel, nor
suborning of witnesses could save us, and it would be of no
use to plead "not guilty," when we were guilty. If He be
just and good, then we know, that in order to have any
sympathy or communion with Him, we too must be just and
good. What He loves, we must love ; what His holy nature
repels, we must repel ; or else we array ouselves in perpetual
warfare against Him ; and such a warfare must be fatal to the
weaker party—that is to us.

Since, then, there must be the most intimate relation, a relation at all points between God and ourselves, no fact can be
more important for us to know than what that relation is. In
what relation then do we stand to our Creator, Governor, and
final Disposer? And in what correlative relation does He
stand to us? The grand and paramount relation in which God
stands, not only to the human race, but to all creatures that
have life, and to all things that have no life, is that of Law-

giver. He has made laws for whatever he has created. There
is not a constellation so vast, nor an atom so small; there is not
an arch-angel so exalted, nor an animalcule so insignificant,
that is not all penetrated and encompassed, bound in and
bound down, by the law which God has impressed upon
its being. In us no thought, nor desire, rises into the light of
consciousness, no muscle contracts, no nerve vibrates, no drop
of blood flows in our veins, or hair grows upon our heads, but
in accordance with the laws which God has severally impressed
upon them. But before going further, let us understand the
meaning of this superb and majestic word, Law.

"Law," says Mr. Justice Blackstone, "in its most general
and comprehensive sense, signifies a rule of action; and is
applied indiscriminately to all kinds of action; whether animate
or inanimate, rational or irrational. Thus we say, the laws of
motion, of gravitation, of optics, or mechanics, as well as the
aws of nations." Thus, when the Supreme Being formed
the universe, and created matter out of nothing, He impressed
certain *principles* upon that matter from which it never can
depart, and without which it would cease to be. When He
put that matter into motion, He established certain *laws of
motion*, to which all movable bodies must conform. If we
further advance from mere inactive matter to vegetable and
animal life, we shall find them still governed by laws, more
numerous indeed, but equally fixed and invariable. This,
then, is the general signification of the word law—a rule of
action dictated by some Supreme Being. Now let us look at
some of the characteristics of these laws:

1st. They are *resistless;* that is, God is resistless in whatever
he does. If He has attached pain and early decrepitude to
gluttony and intemperance, can you detach them so as to con-
nect the offense and still escape the penalty? When He has

made the air a necessity for the lungs, can you fill them with water and still live?

2d. They are uniform.

3d. They are so *comprehensive*, so *universal*, that they absolutely exclude all *chance*.

4th. They are all accompanied by a *sanction*; that is, there is a penalty that attends their transgression. No human law can execute itself. But God's laws execute themselves. God forbids intemperance, and what unutterable woes come to avenge it. God commands cleanliness, and if people will wallow in filth and imbibe foulness at every pore, what pestilences stalk forth to avenge his violated law! From human penalties men sometime escape; they cancel the offense, they foil the prosecutor, they flee the country; but who can secrete anything from the All Seeing Eye? who can circumvent the All Knowing Mind? who can flee beyond the jurisdiction of the Omnipotent? For all conceivable, for all possible offenses against God's laws, whether of the body or of the mind, there is an inexorable, an irrepealable law, that the offenders must suffer until the offenses cease.

INDEX.

INDEX.